Subterranea

Subterranea

DISCOVERING THE EARTH'S
EXTRAORDINARY HIDDEN DEPTHS

Chris Fitch

WITH MATTHEW YOUNG

WILDFIRE

Copyright © 2020 Chris Fitch
Maps copyright © 2020 Matthew Young

The right of Chris Fitch to be identified as the
Author of the Work has been asserted by him in
accordance with the Copyright, Designs and
Patents Act 1988.

First published in 2020 by WILDFIRE an
imprint of HEADLINE PUBLISHING GROUP

1

Apart from any use permitted under UK
copyright law, this publication may only be
reproduced, stored, or transmitted, in any form,
or by any means, with prior permission in
writing of the publishers or, in the case of
reprographic production, in accordance with
the terms of licences issued by the Copyright
Licensing Agency.

Every effort has been made to fulfil
requirements with regard to reproducing
copyright material. The author and publisher
will be glad to rectify any omissions at the
earliest opportunity.

Cataloguing in Publication Data is
available from the British Library

Hardback ISBN 9781472272324

Designed by Matthew Young

Printed and bound in Italy by LEGO S.p.A.

Headline's policy is to use papers that
are natural, renewable and recyclable
products and made from wood grown
in well-managed forests and other
controlled sources. The logging and
manufacturing processes are expected
to conform to the environmental
regulations of the country of origin.

HEADLINE PUBLISHING GROUP
An Hachette UK Company
Carmelite House
50 Victoria Embankment
London EC4Y 0DZ

www.headline.co.uk
www.hachette.co.uk

For Mum, Dad and Charlotte,
my support crew

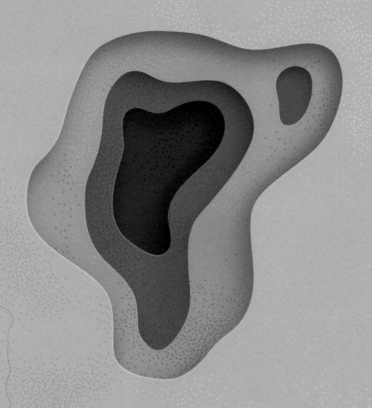

PART III: MODERN HISTORY

PART IV: TODAY

Introduction

In 1691, a young geophysicist with shoulder-length hair made a presentation to the Royal Society of London that was, even by the standards of seventeenth-century breakthroughs, fairly astounding. The scientific community had been flummoxed by the unpredictability of Earth's magnetic field, by the way the planet's poles consistently shift over time. To Edmond Halley — he of the famous comet — the solution to this conundrum was simple: inside the planet must be a series of concentric inner worlds, each separated by gravity. The ground on which we stand is therefore simply the 500-mile-thick outermost layer, and it's the movement of these internal worlds that keeps throwing off the magnetic readings. Furthermore, he predicted that these worlds are inhabited by life forms, and lit by an unknown subterranean light that also explains the existence of aurora borealis, the Northern Lights.

Halley wasn't necessarily plucking these theories from thin air. Instead, he leant heavily on millennia of speculation about what lay beneath our feet. From ancient mythology through religious notions of a punishing afterlife, humanity has repeatedly theorised about fantastical underworlds, often as a counterweight to the glorious heavens above. Dante's escape from Hell in his infamous *Inferno* (*The Divine Comedy*) brought one such vision to life in hauntingly graphic detail. But the credentials of a great man of science like Halley gave the theory of underground worlds real empirical weight. Halley was so committed to this theory that in his final portrait — in 1736, aged eighty — he is shown clutching a parchment that clearly depicts a diagram of these inner layers.

Sadly for Halley, the scientific community had little time for what has since become known as the 'hollow earth' hypothesis. Subsequent experimentation established there to be significant density to the planet's interior, as illustrated by measurements of seismic waves passing through the upper and lower mantles that are now widely accepted to exist beneath the continental crusts upon which we walk. A series of layers, yes, but ones comprised of liquid magma and a solid iron inner core, instead of prehistoric animals inhabiting their own mini-worlds.

And yet. No human has ever set eyes upon a core, or seen beyond the merest glimpse of a mantle. The deepest hole in the world — the Kola Superdeep Borehole in Russia — has penetrated only 12 kilometres (7 miles) into the ground, a fraction of the roughly 6,350-kilometre (3,945-mile) radius of the Earth. So it's hardly surprising that vivid imaginations have continued to dream up their own realities, spawning endless theories and tall tales about deep and mysterious subterranean worlds.

Fiction writers have certainly found their imaginations stirred creatively by the prospect of new worlds below our own, so much so that it has essentially evolved into its own genre. Perhaps the most famous is Jules Verne and his typically adventurous *Journey to the Centre of the Earth*, where terrifying monsters battle to the death in a vast underground ocean. But subterranea has been the gift that keeps on giving, from Edgar Rice Burroughs's Pellucidar to the psychedelic Wonderland found by Alice when she went tumbling down the rabbit hole. Conspiracy theorists have happily got in on the act, arguing that the existence of a vast, secret, far superior subterranean world is perhaps the greatest cover-up of all time.

Certainly, with a surface world that has been explored, surveyed, measured, mapped, photographed and Instagrammed to the point of sterility, perhaps the subterranean realm is where Earth's remaining mysteries — natural and anthropogenic — are

yet to be found. It's certainly worth acknowledging the limitations of the horizontal plane most of our lives takes place in. We do in fact live in a world of (at least) three dimensions, and while our instinct is often to look up, sometimes the best stories can be found by looking down. If you were to peel back the Earth's surface like an orange, then take a sly peek underneath, what remarkable things would you see?

In *Subterranea*, we'll explore caves full of tiny blind dragons, and modern mass-transit networks that uncover the very bones of history on which our cities are built. Tunnels built as protection from the brutality and fear that comes from conflict, as well as the great optimism of people storing precious treasures underground, safe for an uncertain future. From enigmatic cave drawings to nuclear bunkers, talking trees to futuristic underground infrastructure, subterranea remains just as evocative and awe-inspiring as it was in Halley's day. It's time for the strange and uncomfortable realities of a world we perhaps don't know as well as we think we do to be abruptly exposed to the light.

Photographs on preceding pages:

Creation

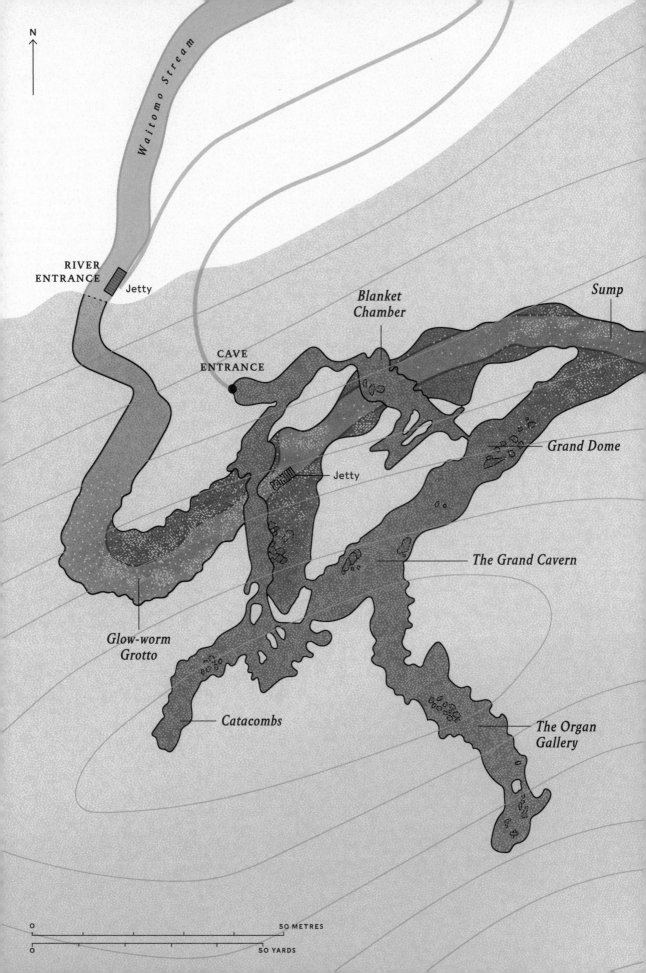

N

Waitomo Stream

RIVER
ENTRANCE
Jetty

Blanket
Chamber

Sump

CAVE
ENTRANCE

Grand Dome

Jetty

The Grand Cavern

Glow-worm
Grotto

Catacombs

The Organ
Gallery

0   50 METRES

0   50 YARDS

Waitomo Caves

WHERE MILLIONS OF GLOW-WORMS CREATE A CONSTELLATION OF LIGHTS

NEW ZEALAND

S 38° 15′ 40″
E 175° 06′ 15″

A dizzying spectacle, as though the richest of night skies has been transported underground, subverting our understanding of physics and astrology. This subterranean starlight spreads across the cave ceiling, with brief bare patches appearing to the uncomprehending eye to resemble silhouettes of tree canopies blocking out starlight overhead. However, this luminescence comes not from distant stars, but instead from organisms. Extraordinarily, every single spot of light betrays the presence of an individual animal — one found almost nowhere else on Earth.

In the South Pacific, layers of limestone were formed beneath the sea over 30 million years from the remains of ancient marine organisms, creating rock packed full of shell and skeleton fossils from these long-deceased creatures. Thanks to the immense forces of geology and volcanic activity, these layers began to buckle and bend. Once exposed to the elements above the surface of the ocean, they slowly started breaking apart, creating small cracks and weaknesses in the rock. Eventually, some cracks — such as in the western region of what is now New Zealand's North Island — grew sufficiently large to become what we might consider proper caves.

These subterranean spaces remained untouched by human hands and untrodden by human feet for centuries, even after the first arrivals of Māori settlers in the land they call Aotearoa ('land of the long white cloud') around the year 1200. In a similar vein to the mountains and rivers that dominated this rugged landscape, caves were seen as sacred. They were the realm of the dead, and therefore stringently avoided. The only thing people knew about the roughly 300 caves that dotted this particular region of the country was that various channels of water seemed to disappear into the ground, hence the name gifted to the area: Waitomo (*wai* meaning 'water' and *tomo* meaning 'entrance' or 'hole').

Nevertheless, these fierce warriors never dared to venture into the inky depths below their feet. It wasn't until the late nineteenth century, long after Europeans had begun pouring into the country they now called New Zealand, that anyone decided to peek inside the caves, to see exactly what was down there. English surveyor Frederick Mace was the first to take the plunge, in December 1887, convincing local chief Tane Tinorau to join him on a daring ride aboard a flax-stem raft through an underground river that until then had never had to contend with the presence of humans. 'On seeing that the raft was perfectly safe, Tane Tinorau screwed up his courage and agreed to accompany me,' Mace later wrote in the local *King Country Chronicle*. 'We lit our candles and started upon our voyage of discovery.'

This shortage of lighting wouldn't have been a problem for long. With their jaws surely agape, Mace and Tinorau would become the first humans in history to set eyes upon the glorious constellations that decorated the roof of this cave. Thousands upon thousands of tiny pinpricks of light, combining to create an astonishing display across walls and ceiling alike.

The creature responsible for this incredible light show is *Arachnocampa luminosa*, commonly known as the New Zealand glow-worm, each of which is equipped with a tiny blue-green light. It's the larvae of a minuscule fly called a fungus gnat, not dissimilar in appearance to a mosquito. Peculiarly, these glow-worms operate on a circadian rhythm, their glowing shifting in accordance with the passing of day and night, despite never actually seeing any sunlight during their short lives. For nine months, this Milky Way of maggots stretches across the cave ceiling, seducing

New Zealand's Waitomo Caves are illuminated by the light of thousands of glow-worms, all trying to lure in a meal.

→

nearby insects to approach for a closer look. Once lured in, their prey will almost inevitably become entangled in one of the hundreds of thin, sticky threads that the glow-worms leave dangling from the ceiling. Once snared, the insects can do little except wait to be pulled up by the hungry larvae, then eaten alive. Their only consolation is perhaps the knowledge that, once these maggots pupate into flies, they will have no mouths, and so will live for only a few days of intense procreation before quickly starving to death.

← *The dramatic interior shapes at Waitomo have slowly formed over the past 30 million years.*

Glow-worms aside, there is an abundance of troglobite life within the hugely enigmatic rock formations in this cavern, including eels, beetles, spiders and wetas (large, cricket-like insects). Dangling stalactites beg to be touched, but giving into temptation is forbidden, since natural oils in human skin will damage them. Some stalactites are thin and delicate, like they could snap at any moment; some resemble a wave, a curtain, or honeycomb; others are thick and sturdy as tree trunks, resembling limestone pillars or melting wax. Some emerge from the bottom of bulbous rocks like teats on an udder.

The Waitomo Caves also give a brief glimpse into the life of another species unique to this corner of the South Pacific: the moa. These huge birds from the ratite family, similar in appearance to overgrown ostrich, were hunted to extinction by early Māori settlers, leaving their decayed remains as the only evidence of what they were like. And there are plenty of such remains in Waitomo. Moa had an unfortunate habit of falling through holes in the earth and meeting a premature end below ground. Excavations of the caves have found rich supplies of moa bones littering the interiors, providing a valuable insight into this species prior to its eradication from the only islands it ever called home.

Mace and Tinorau returned to the cave many times over the following year, and the site soon became a tourist attraction of immense popularity. Early tours lasted up to seven hours, and lucky participants were often given a broken-off piece of stalactite to take home with them. Eventually, guides realised how slowly these rock formations regrew — something in the region of one cubic centimetre every century — and quickly halted the practice. Instead, contemporary visitors take home with them only photographs of remarkable cave features and memories of extraordinary celestial illuminations from deep underground.

Yucatán Cenotes

THE ENTRANCE TO THE MAYAN UNDERWORLD

MEXICO

N 20° 53' 49"
W 89° 14' 20"

As the Mayans were well aware, huge subterranean freshwater stores can be found inside Mexico's cenotes.
→

The ancient Mayan civilisation lasted for at least 3,000 years. The Mayans built over forty large-scale cities across Mesoamerica — encompassing much of modern Mexico, Guatemala and Belize — and had a peak population of around 2 million people. They used sophisticated agricultural methods such as irrigation, invented advanced architectural techniques, and successfully manufactured resources such as chocolate, rubber and paper. Yet their environment was arid and harsh. The survival of Mayan society, relatively developed as it may have been, hinged on a vast underground world of water, accessible through thousands of gaping holes in the ground.

These cenotes, as they are called, can still be seen today, primarily on the Yucatán peninsula of southeast Mexico. Two such holes can be found next to perhaps Mexico's most famous ancient Mayan landmark, the UNESCO World Heritage site of Chichén Itzá. The name itself tells a rich story about the cenotes, bringing together *chi* (meaning 'mouth'), *chen* ('wells') and Itzá, the name of the specific tribe that settled there. 'Wells' gives a clear indication of why cenotes were so vital to the people living here, as these natural wells were their sole source of drinking water. Each impressive Mayan city on the Yucatán is situated next to at least one cenote, almost certainly its most precious resource. Without cenotes, Mayan civilisation would likely have ground to a halt before ever getting started.

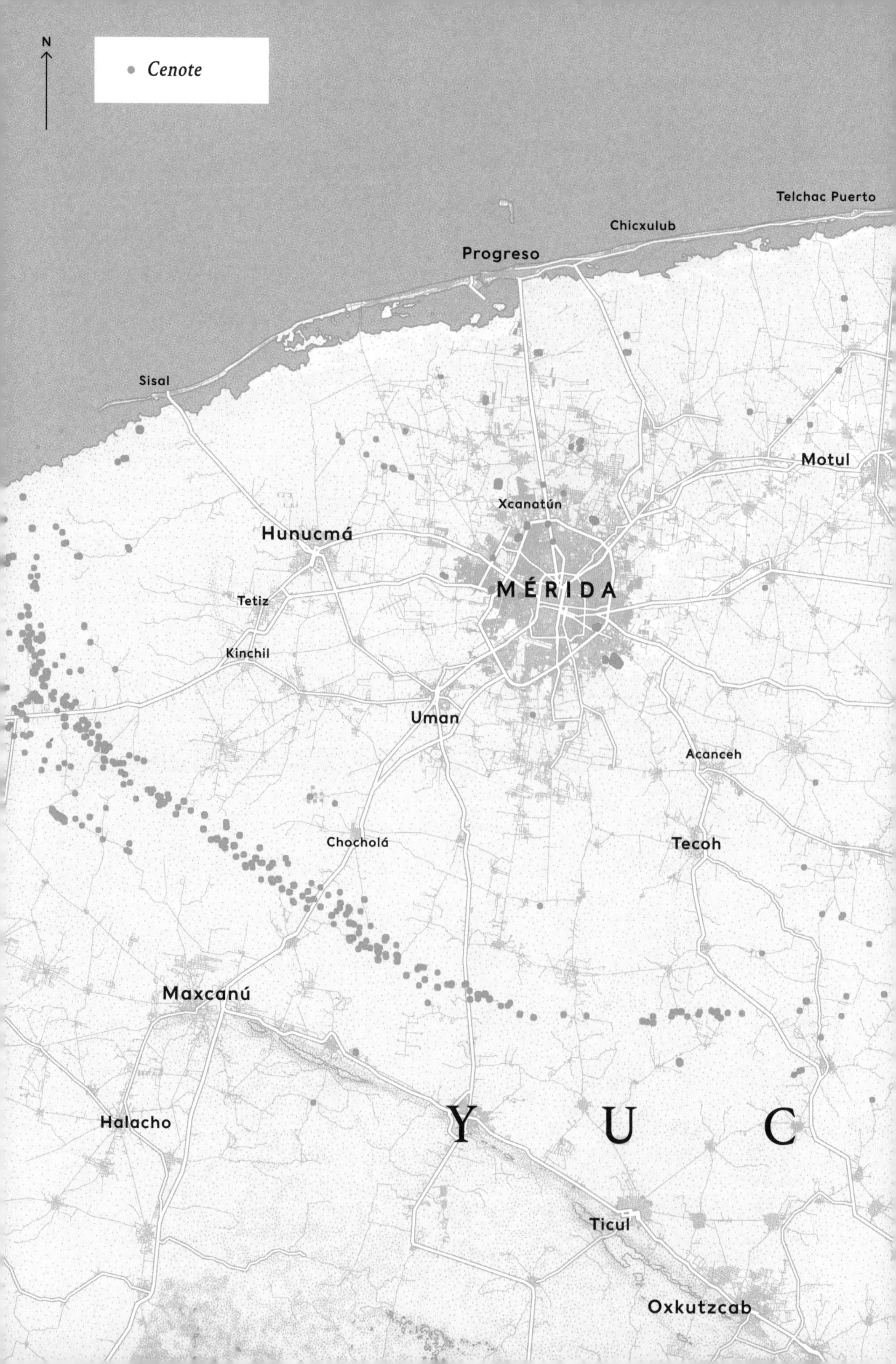
N
Cenote
Telchac Puerto
Chicxulub
Progreso
Sisal
Motul
Xcanatún
Hunucmá
MÉRIDA
Tetiz
Kinchil
Uman
Acanceh
Chocholá
Tecoh
Maxcanú
Halacho
Y U C
Ticul
Oxkutzcab

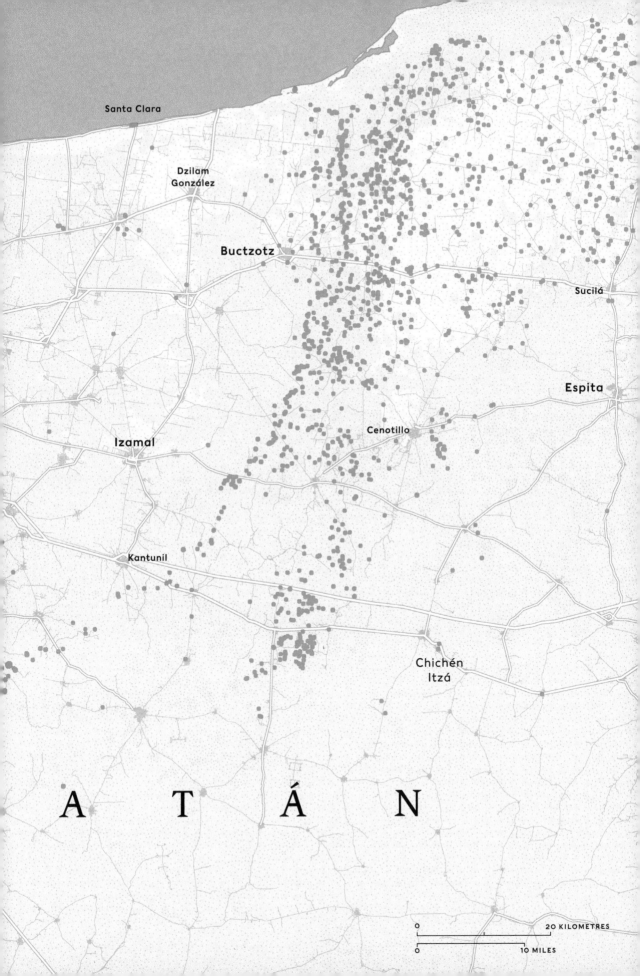

Santa Clara

Dzilam
González

Buctzotz

Sucilá

Espita

Izamal

Cenotillo

Kantunil

Chichén
Itzá

A T Á N

0   20 KILOMETRES

0   10 MILES

↑
There are hundreds of
cenotes spread across
the Yucatán peninsula,
a distinctive feature of
the landscape.

But these holes in the ground, pits across a pockmarked surface, also had a more sinister purpose. Traditionally, they were used to make sacrifices to the god of rain, Chac (also sometimes spelt Chaac or Chaahk) — often depicted in statues with large fangs, and sometimes a snout. Precious offerings would be thrown into the cenotes, in the hope he would respond by bringing the vital seasonal rains necessary to make crops grow. These gifts included everything from jade and gold to incense and even, disturbingly, small children. It is from these offerings that the 'mouth' part of the name arises. Folklore also tells us that cenotes were sacred passageways that led to a netherworld known as Xibalba. An ancient Mayan scripture titled the Popol Vuh describes the truly gruesome path that led to Xibalba, which involved crossing rivers filled with scorpions, blood and pus.

With the aid of modern academia, we now know that cenotes are essentially a specific, isolated form of sinkhole, one that formed when the limestone surface level collapsed, exposing vast cavities underneath. But a deeper theory explaining their creation takes us back to the end of the age of dinosaurs. In northern Yucatán, the town of Chicxulub is where an estimated 15-kilometre-wide (9-mile) asteroid smashed into the planet 66 million years ago, causing a super-tsunami that swept the globe, killing up to 80 per cent of life on Earth. Once the clouds of sulphur had eventually cleared, a huge crater was revealed, up to 180 kilometres (112 miles) in diameter. Half of it is buried offshore under 600 metres (nearly 2,000 feet) of sediment, but the other half, now on land, was later covered by a layer of limestone. It is that layer that has worn away over time, creating the cenotes we know today.

Clear reinforcing evidence for this theory exists in the form of a long line of cenotes that, when viewed from above, stretch across the landscape in an arching semi-circle precisely where the expected outer ring of an impact crater would be. Ninety-nine such cenotes have been officially classified on this arch, with as many as 900 unofficially recognised as part of the same geological feature. The collapse of these cavernous holes in the surface revealed the water table below, where a mix of salt- and freshwater flows congregate. Many cenotes can be considered the starting point for subterranean rivers that flow, eventually, to the sea.

The scale of the space beneath the surface is epic. Hundreds of miles of underground waterways have already been mapped, and ongoing surveying work is far from being complete. In 2020, the vast Sistema Sac Actun submerged network of channels, connecting various cenotes across the Yucatán, was measured at a massive 347 kilometres (216 miles), making it the world's largest underwater cave system.

With such a huge area hidden since antiquity, it's perhaps no surprise that there are many secrets still being covered by speleologists. One cenote, called Xoc, near the city of Mérida in northwest Yucatán, was found to contain remnants of many ancient marine creatures, including thirteen teeth believed to have belonged to the gigantic megalodon shark — quite possibly the largest shark that ever lived — many millions of years ago. Another, Chan Hol near Tulúm, in the neighbouring state of Quintana Roo, yielded a human skeleton dated (by measuring calcite growth) to over 13,000 years old, among the oldest known evidence of a human presence in the Americas.

The Mayan god Chac demands sacrifices in exchange for sending the seasonal rains.
↓

More recent discoveries in the cenotes have been of the modern variety, principally rubbish and human waste. Yucatán's Secretariat of Sustainable Development has estimated that 60 per cent of the state's 2,241 registered cenotes have problems with pollution. In recent years, local teams of divers have been organised to leap into selectively chosen cenotes and retrieve hundreds of tons of waste, including car tyres, glass bottles and old bicycles, as well as a random assortment of items reportedly used in 'witchcraft'. They use a carefully concocted cocktail of bacteria, fungi and microorganisms to biochemically clean these contaminated cenotes. In this way, they can restore the clean water required for a healthy natural ecosystem — which would surely please Chac even more than sacrificial jewels or children.

Postojna Cave

SLOVENIA

N 45° 46' 59"
E 14° 12' 06"

In 1818, Postojna Cave, situated within the dominating Austrian Empire, consisted of a chamber a few hundred metres deep that was attracting curious tourists from afar. In April, the locals were preparing for the grand arrival of Austrian Emperor Francis I (simultaneously Francis II of the Holy Roman Empire) and his fourth wife, Caroline Augusta. They were keen to see the impressive spectacle as part of their journey on a visit to Dalmatia, in modern-day southern Croatia. As the cave crew prepared for the visit, a low-level assistant lamplighter named Luka Čeč was given the dangerous assignment of crossing a subterranean river on a makeshift bridge to affix a welcome sign.

Upon reaching the other side, Čeč disappeared out of sight beyond what was believed to be an impenetrable rocky barrier. Onlookers were left concerned about what had happened to him, whether the earth had swallowed him up, never to return. When Čeč eventually re-emerged, his face gleaming with excitement, he claimed to have discovered 'Paradise' deep within the very heart of the cave. To corroborate his story, he showed pieces of stalactites and stalagmites he had broken off in order to mark his path back to the entrance.

It was a moment that would forever transform this region of what has since become modern Slovenia. Postojna is now known to be a remarkable place full of enormous cathedral-like spaces,

N

The Pivka River
disappears for 2,200m,
resurfacing at Planina

Pivka Cave

Pivka River

siphon

Black Cave

Magdalene's Cave

ENTRANCE

siphon

siphon

siphon

Bertarelli's Passage

Paradise Cave

Calvary
(Great Mountain)

Tricolor Cave

ENTRANCE

Island Cave

Diamond
Hall

Iris
Passage

The
Curtain

Pivka River

Railway

Veliki Otok

RIVER ENTRANCE

CAVE ENTRANCE

Postojna

0   500 METRES

0   500 YARDS

Pivka River

with columns as tall as buildings, including what is affectionately known as the 'skyscraper', a grand 16-metre (52-feet) stalagmite that has been dated to around 150,000 years old. With 24 kilometres (15 miles) of underground paths, it is Europe's longest cave open for public viewing.

Yet relatively little is known about Čeč, the man who discovered the true depths of Postojna. There are no known photos of him. He tried to claim some mild fame for his remarkable discovery; despite a poor background and low social status, he appealed to the emperor, requesting to be given due credit. But his request was denied, and the accomplishment instead granted to Josip Jeršinović von Löwengreif, district treasurer of the Postojna region. Only decades later was Čeč's vital role acknowledged, and he was subsequently officially recognised as the cave's true discoverer — unfortunately, many years after the man himself had passed away.

Čeč was also the first to find what is now known as the slender-necked beetle, a spindly copper-coloured insect with a slim thorax (upper body) and a proportionally enormous abdomen. The study of this beetle launched the entire academic discipline of speleobiology (also known as biospeleology) — the study of cave animals. But of the 150 different animal species that have been found in the depths of Postojna, the main attraction is undoubtedly a peculiar amphibian called *Proteus anguinus*, or 'olm'. In local medieval folktales, these creatures were called 'baby dragons'. These are not the flying, fire-breathing beasts of myth, legend and Hollywood blockbusters. These 'dragons' are considerably slower and calmer, although no less fascinating. They might in fact be one of the world's most peculiar amphibians.

Olms are one of the world's strangest amphibians, able to live for up to twelve years without eating.
→

Olms are Europe's only cave-adapted vertebrates, inhabiting subterranean aquatic systems beneath the Dinaric Alps, which straddle Slovenia, Croatia, Bosnia and Herzegovina and beyond. These curious animals resemble small, pale, almost translucent salamanders, and are completely blind; they do have eyes, but they are so weak as to be almost useless. They employ a powerful sensory combination of smell, taste, hearing, electromagnatism and photosensitive skin (they can essentially *feel* light) to find food, in the form of small crabs, worms and snails. Not that they are frequently in desperate need of a meal; these creatures can go for up to twelve years without eating, thanks to their ability to stay almost completely still, using practically no energy whatsoever.

An estimated 4,000 olms live in Postojna, where they are forensically studied for clues to explain their extremely long and drawn-out procreation cycle. The laying of olm eggs takes around six or seven years. White, jelly-coated pearls are attached to the walls of the cave, before being individually fertilised. Once born, the lifespan of an olm is similarly elongated, with some living long enough to become centenarians. Despite the extreme oddity and general lack of charisma of these alien-like organisms, Slovenians hold a quaint fondness for them, even immortalising them on a national coin. From remarkable cave formations to truly bizarre fauna, Postojna can lay claim to being one of the most extraordinary subterranean spaces in Europe. Čeč would surely have been proud.

Upon returning from his first delve into Postojna, Luka Čeč reportedly told colleagues he had discovered 'Paradise'.
→

Veryovkina Cave

POTENTIALLY THE WORLD'S DEEPEST CAVE

GEORGIA

N 43° 24' 56"
E 40° 21' 23"

Deep underground, a team of cave explorers races against time, clambering to escape rising floodwaters. Half an hour earlier, a warning had been sent from teammates nearer to the surface, notifying their colleagues of heavy rainfall from a passing storm. But the threatening reality of the situation would only sink in when, mid-breakfast, the explorers below could hear for themselves the pulsing sound of approaching water, a deafening roar that chilled them to the bone. Grabbing climbing gear, sleeping bags and memory cards, they had no choice but to abandon the rest of their belongings and flee, in a desperate rush towards the surface, over a mile above their heads.

The title of 'world's deepest cave' is more hotly contested than might be assumed. Caves are complex, disorganised places. Locating their deepest point is profoundly more difficult than finding the highest point of a mountain. Tunnels can splinter, fracture, branch out into a multiple of possible paths. Openings might be easily hidden from sight, or be visible yet completely out of reach. Sometimes a dead end may appear, only for a tiny gap mere inches wide to lead to a vast chamber or ongoing passageway beyond. Extreme darkness and cold, plus the threatening and unpredictable presence of underground rivers, further complicates the exploration process.

For years, speleologists had wrestled over the depth of Krubera, a cave in the Abkhazia region of Georgia, a rugged part of the world

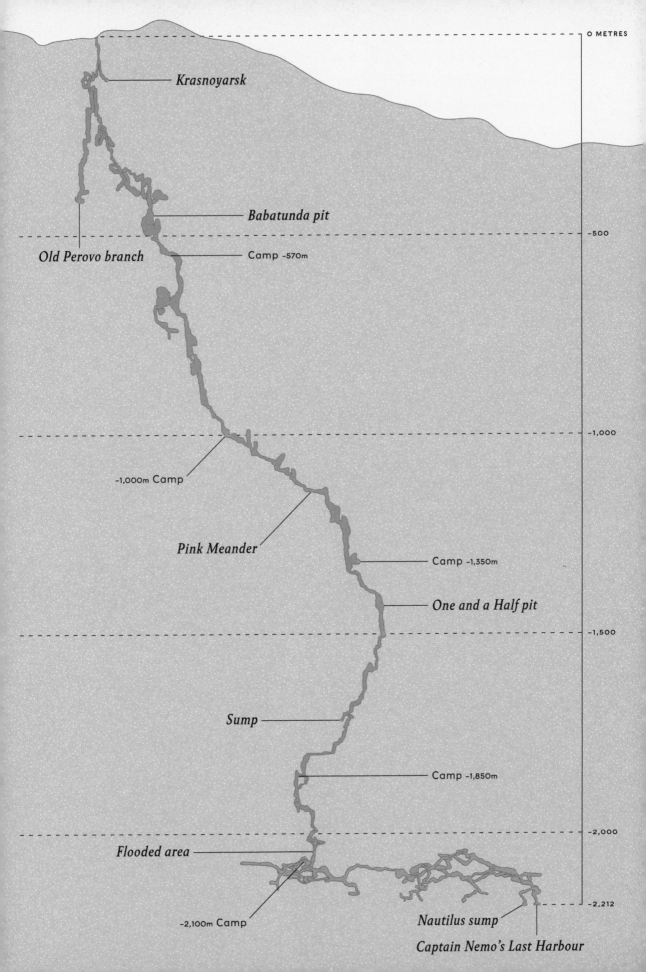

O METRES

Krasnoyarsk

Babatunda pit

Old Perovo branch

Camp –570m

–500

–1,000

–1,000m Camp

Pink Meander

Camp –1,350m

One and a Half pit

–1,500

Sump

Camp –1,850m

–2,000

Flooded area

–2,100m Camp

Nautilus sump

Captain Nemo's Last Harbour

–2,212

that lies between the Greater Caucasus Mountains and the Black Sea. Successive expeditions since 2001 had established the cave to be the world's deepest, first at least 5,000 feet below the surface, then 6,000 feet, then 7,000 feet. But perhaps the greatest breakthrough came in 2012, when Ukrainian cave explorer Gennady Samokhin attempted to crawl and dive his way deeper than anyone had been before. Beginning on a high plateau 2,240 metres (7,349 feet) above sea level, he battled subterranean spiders and other critters, pitch-black submerged tunnels and freezing temperatures, in order to establish a new official depth of at least 2,197 metres (7,208 feet). Krubera was now surely, undisputedly, the world's deepest cave.

But only a few mountains away, a challenger was waiting. While Samokhin was speaking to the press about his belief that Krubera continued even deeper into the Earth, a new expedition was eyeing up nearby Veryovkina Cave. In March 2018, a Russian-led team travelled to Abkhazia and spent nearly two weeks inside Veryovkina — entering through a small shaft in the mountainside 2,285 metres (7,497 feet) above sea level — to see how far down this cave went. The interior was damp and tortuously discombobulating, with dead ends and tight squeezes through tiny cracks at every turn. As they approached depths comparable with the most extreme in Krubera, they stumbled upon a large lake, pitch-black, glassy still, forcing the rest of the expedition to be conducted underwater. Eventually, cave divers resurfaced to confirm the news that everyone had been waiting to hear: the absolute depth of Veryovkina exceeded that of Krubera, albeit by only 15 metres (50 feet). It was incredibly close, but a new champion had been crowned.

Later that year, a new expedition headed back to the same location, to delve deeper into the mountain and see how much further down it might go. After many rare and newly discovered troglofauna, such as shrimp and scorpions, had been collected on the previous trip, there was also an enthusiasm to add to the academic knowledge about the cave. A fresh team of explorers kitted up and attempted to plunder further into the depths of

↑

Descents into the world's deepest known cave involve perilous abseils through freezing conditions and near-total darkness.

Veryovkina, to see how much deeper the cave could go — complete with a *National Geographic* photographer ready to document the monumental achievement.

But when rapidly rising floodwaters threatened the expedition (making a mockery of their belief that the cave only flooded in wintertime), they were forced into a desperate scramble for their lives. Frustratingly for these explorers — who had already spent a week underground — they were left with no option but to abandon this quest and return to the surface world. To date, Veryovkina's official depth remains 2,212 metres (7,257 feet) below the surface, a world record that waits, tantalisingly, for someone to break it again.

Kazumura Cave

AN EXCEPTIONALLY LONG LAVA TUBE,
STEEPED IN MYTHOLOGY

HAWAII, USA

N 19° 29' 15"
W 155° 04' 47"

In Hawaii, legend tells us of two sisters arriving from Tahiti, and settling on the island. The elder, volcano goddess Pelehonuamea (also known as Pele), having recently fought with one powerful sister (Namakaokaha'i, goddess of the sea), soon falls out with another, her companion younger sibling, Hi'iaka'aikapoliopele (thankfully commonly shortened to Hi'iaka). After a night in which she is captivated by the dashing young chief Lohi'au, Pele orders Hi'iaka to bring this man to her home, in the crater in the highlands of the island, within forty days. Failure, she tells her sister, will see her destroy a forest on the island of which Hi'iaka is especially proud.

Hi'iaka undertakes an epic journey, and is ultimately successful. However, she takes so long to complete the task that Pele, apparently not a patient sort, assumes her sister has taken Lohi'au for herself, and proceeds to pummel Hi'iaka's forest with fire, burning it to the ground. When Hi'iaka retaliates for this act of aggression, Pele responds with rivers of lava, killing Lohi'au, and sending his body tumbling into the crater. Hi'iaka tries desperately to retrieve Lohi'au's body, digging furiously into the flaming pit. Overall there's a lot of fire, flying rocks and general volcanic melodrama.

This story, and many like it, explain not only the fiery volcanic landscape of Hawaii and surrounding islands, but many of the epic eruptions that have taken place throughout the millennium or so that humans have inhabited the islands. Folklore tracks all

Kazumura is now known to be at least five times longer than when it was first recognised as the world's longest lava tube.
→

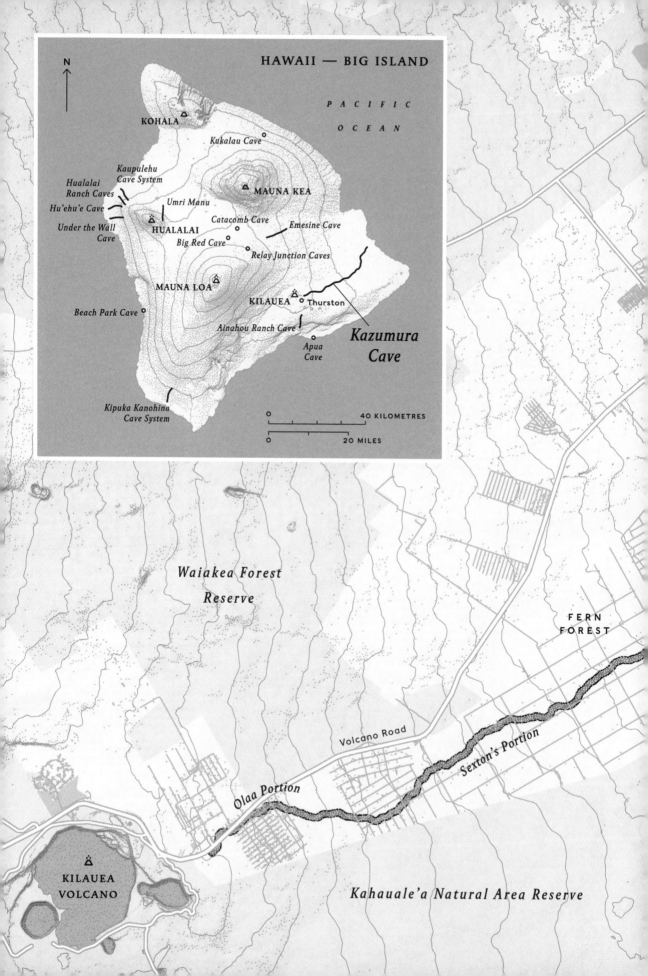

HAWAII — BIG ISLAND

N

KOHALA

PACIFIC

OCEAN

Kukalau Cave

Kaupulehu
Cave System

MAUNA KEA

Hualalai
Ranch Caves

Umri Manu

Hu'ehu'e Cave

Catacomb Cave

Emesine Cave

Under the Wall
Cave

HUALALAI

Big Red Cave

Relay Junction Caves

MAUNA LOA

KILAUEA

Thurston

Beach Park Cave

Ainahou Ranch Cave

Kazumura
Cave

Apua
Cave

Kipuka Kanohina
Cave System

40 KILOMETRES

20 MILES

Waiakea Forest
Reserve

FERN
FOREST

Volcano Road

Sexton's Portion

Olaa Portion

KILAUEA
VOLCANO

Kahauale'a Natural Area Reserve

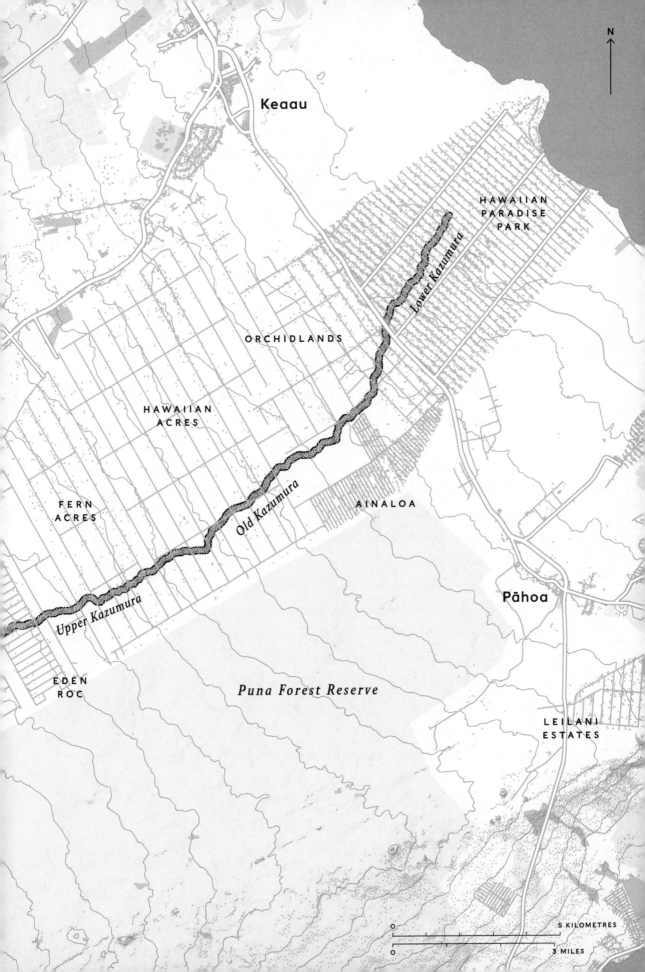

Keaau

HAWAIIAN
PARADISE
PARK

Lower Kazumura

ORCHIDLANDS

HAWAIIAN
ACRES

FERN
ACRES

Old Kazumura

AINALOA

Pāhoa

Upper Kazumura

EDEN
ROC

Puna Forest Reserve

LEILANI
ESTATES

N

0   5 KILOMETRES

0   3 MILES

the big events, from the aforementioned burning of forests, to wild projectiles of rocks caused by Hi'iaka's digging, which geo-mythologists can match with data collected from their geological research to more accurately date past eruptions. Mythology might also help us understand the creation of the Kazumura Cave — the longest and deepest lava tube in the world.

Kazumura was formed by a truly massive flow of lava, during an estimated sixty-year eruption of Mount Kilauea, believed to have occurred sometime in the middle of the fifteenth century. This eruption was named after Aila'au, the fire god who devours forests (as the name roughly translates), who, despite this fearsome reputation, supposedly fled when Pele arrived. As the ongoing eruptions finally died down, the torrent of lava that gushed towards the ocean began to cool from the outside inwards. While hard basalt rock was forming around the outside, lava continued to flow as something close to liquid on the inside. Eventually, once the molten rock river inside had fully drained away, this left a huge natural 'tunnel', stretching all the way from the edge of the Kilauea volcano to near the water's edge.

In the highly active hotspot of Hawaii (where frequent lava flows, low in viscosity, go by the local name of *pāhoehoe*), such tubes are relatively commonplace. Spread out across the landscape, many are shallow and possess hazardous openings to the surface known as *puka*; often covered by vegetation, they can be a trap to unwary passing walkers, who have been known to fall inside, some-times with fatal results. Crucially, such tubes can be created rel-atively quickly, sometimes during a single eruption — far faster than the slow processes of tectonic movement and erosion that form most caves and other geological features.

Despite such ubiquity, Kazumura dwarfs all peers when it comes to scale. It's huge — though establishing the exact length of Kazumura has been a surprisingly difficult undertaking. The official length was once recorded as 11.7 kilometres (just over 7 miles) when first explored in 1981, still enough to make it the world's longest lava tube. But since the first complete walk-through in 1995, it was extended to a whopping 64 kilometres (40 miles),

after it was confirmed that five separate lava tubes were in fact all segments of the same incredibly long tunnel.

There are now over a hundred known entrances to the cave between its starting point on the slopes of Kilauea volcano and where the cave almost meets the ocean, a vertical descent of over a kilometre. The tube is easily large enough to walk through comfortably — perhaps even drive through, were it logistically possible to get vehicles inside — with a ceiling as high as 18 metres (60 feet), and walls up to 21 metres (69 feet) apart.

Discombobulated tree roots protrude sporadically through the ceiling. It's also hard not to be distracted by remarkable displays of lavacicles (perhaps better known as 'lava stalactites'), the remnants of dripping molten rock. The walls are coloured in bright minerals, including dark magnesioferrite and green olivine. It's an otherworldly, alien-like environment, understandably considered *kapu* (sacred) by indigenous Hawaiians. Its presence is befitting for an island whose folklore is dominated by clashes between terrifyingly powerful deities from another world.

Cueva de Villa Luz

ACID-FILLED CAVES IN WHICH
UNIQUE FISH SOMEHOW SURVIVE

MEXICO

N 17° 27' 00"
W 92° 47' 45"

Wading through milky water, a group of men in white shirts, wide-brimmed hats and bright red neckerchiefs enter a cave system. It isn't immediately obvious, but this cave is far more hostile than even the most imaginatively pessimistic claustrophobe would likely imagine. As they do every spring, the Zoque — a community located in south Mexico — begin dipping leaves coated in a paste into the cave water, part of an ancient fertility ritual asking for rains to break the long, dry spell above ground. In response, fish begin floating up to the surface, completely immobilised. The worshippers quickly gather their stunned prey into baskets, thankful for this gift from the gods, a bounty to tide them over until the rains return.

It's an unusual place to be looking for food. Cueva de Villa Luz ('Cave of the Lighted House'), as this underground world is known, is famous for its inhospitable interior conditions. These waters are rich in sulphuric acid, hence the pale colouring of *El Azufre* ('the sulphur') river that flows out onto the surface. Thanks to the presence of microbubbles of hydrogen sulphide that make their way up from buried deposits of oil below ground — producing a thick, pungent scent of rotten eggs in the air — and react with dissolved oxygen in the stream, sulphuric acid soon forms, driving down the pH of the waters. This is supplemented by the significant presence of the somewhat nauseating but vividly named 'snottites'. These bleached, bogey-like stalactite colonies of microbial bacteria hang

from the cave ceiling, oxidising that same atmospheric hydrogen sulphide (in lieu of photosynthesis) to produce super-strong sulphuric acid, which drips into the water below. With an average pH of 1.4, and sometimes as low as zero (the strongest on the scale), this is acid capable of burning right through human skin upon contact.

So extreme are the conditions, cave explorers must wear protective breathing apparatuses to fend off the respiratory dangers of the toxic fumes. Without this additional help, a burning sensation in the throat and lungs would likely warn of an impending threat, while unconsciousness and even death wouldn't be far behind. The Zoque make do without such additional equipment, which is perhaps why their ceremony takes place only just

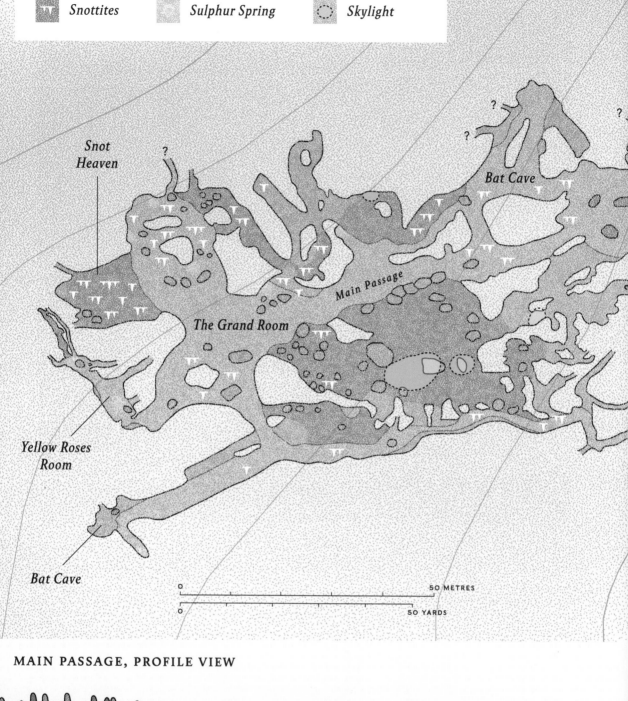

Snottites   Sulphur Spring   Skylight

Snot Heaven

Bat Cave

Main Passage

The Grand Room

Yellow Roses Room

Bat Cave

0   50 METRES

0   50 YARDS

MAIN PASSAGE, PROFILE VIEW

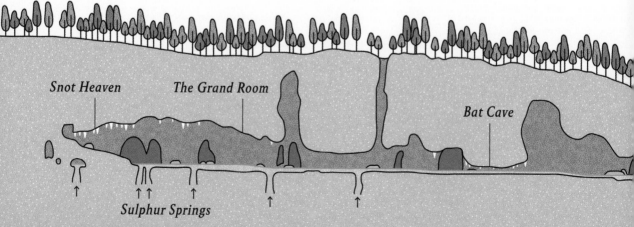

Snot Heaven   The Grand Room

Bat Cave

Sulphur Springs

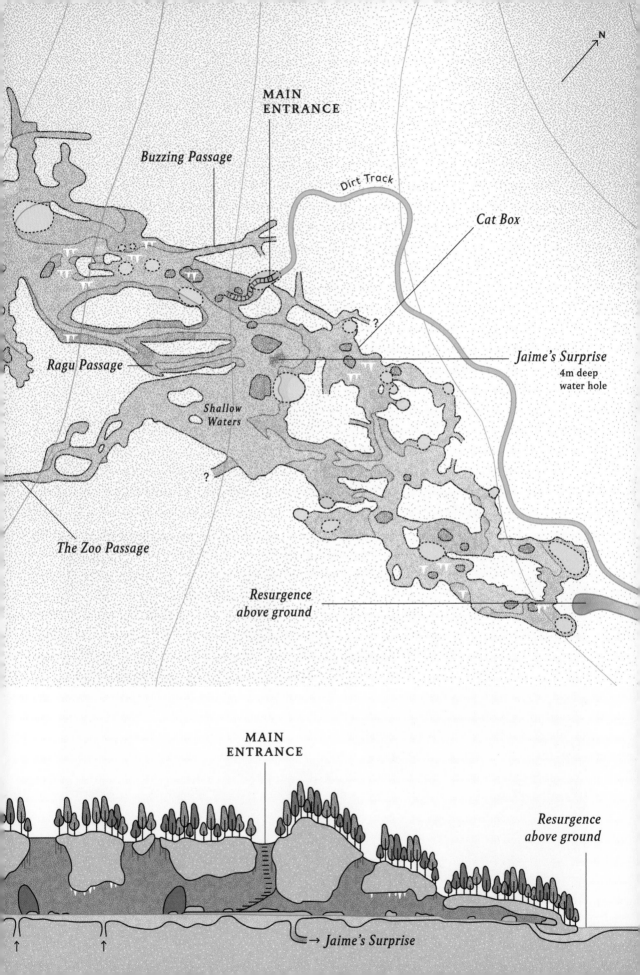

N

MAIN
ENTRANCE

Buzzing Passage

Dirt Track

Cat Box

Jaime's Surprise
4m deep
water hole

Ragu Passage

*Shallow
Waters*

?

?

The Zoo Passage

*Resurgence
above ground*

MAIN
ENTRANCE

*Resurgence
above ground*

→ *Jaime's Surprise*

inside the cave entrance, and is an experience they suffer through only once a year.

It might be assumed that such an environment would therefore be entirely devoid of life. And yet, the Zoque ritual provides pretty irrefutable evidence that even these intense conditions aren't enough to prevent some fish species from inhabiting these caves in their thousands. Atlantic mollies (*Poecilia mexicana*), found in many streams in this part of the world, would not normally be considered remotely capable of surviving in such acidic underground waters. But the population in Cueva de Villa Luz appears to have adopted various biological and behavioural adaptations that make such an unlikely outcome possible. These include so-called aquatic surface respiration — regularly heading to near the water surface to breathe — which reduces the quantity of toxins they consume, mitigating the worst effects. It is even believed they have gained the curious ability to detoxify the acid once it enters their bodies.

The arrival of the Zoque, however, is an event against which even the tenacious mollies are unable to defend themselves. When

→

Colonies of microbial bacteria known as 'snottites' cling to the cave ceiling.

the ceremony is performed, using a paste derived from the ground-up roots of the barbasco plant — a common shrub across the Americas — the poison rotenone seeps into the waters, and is carried through the cave by the currents, anaesthetising the fish. The hunters can then calmly collect their reward, safe in the knowledge that rotenone is only faintly toxic to humans.

This ceremony has proved to be an interesting case study for evolutionary scientists, who have observed how some mollies are more resistant to the root poisons than others, enabling these individuals to breed and pass on their resistance to their offspring. Over time, this has made the population of hardcore cave-dwelling mollies considerably more resistant to the poison than their softer cousins in the outside world. Given their continued resilience in the face of both the toxic environment of the cave and the annual poisonings of the Zoque, these mollies might claim to be among the toughest fish in the world.

Despite the extreme conditions, mollies can be found in the caves in large numbers.

←

Deer Cave

THE BIGGEST UNDERGROUND RIVER PASSAGE IN THE WORLD

MALAYSIA

N 04° 01′ 41″
E 114° 49′ 44″

Steep subterranean cliffs. Huge mounds of guano, covered in bugs. And almost complete darkness. In early August 1977, British explorer Robin Hanbury-Tenison and his teenage daughter were navigating their way through a massive underground river passageway. Ahead, a single bright illumination enticed them towards it, the literal light at the end of the tunnel.

When they had scrambled across the bat droppings and finally reached the exit, emerging into dazzling sunlight, they found themselves in an enclosed valley, surrounded by seemingly impenetrably steep sides all around. This would turn out to be an extraordinary journey in more ways than one: the first recorded climb through what is now believed to be the largest underground river passage in the world, to become the first known visitors to what they later dubbed the 'Garden of Eden'. 'Ours may well have been the first human feet to tread there,' wrote Hanbury-Tenison in *Finding Eden*, his memoir of the expedition.

These words evoke a sensation that was perhaps common to global explorers from the past few centuries. The late 1970s might have been a surprisingly recent time for such an expedition to have occurred, but Mulu, on the southeast Asian island of Borneo, still had significant wonders to reveal. It was six weeks after first arriving that Robin Hanbury-Tenison, Royal Geographic Society expedition leader, undertook this trek through the intimidatingly vast cathedral of Deer Cave. Over a kilometre long, it was so

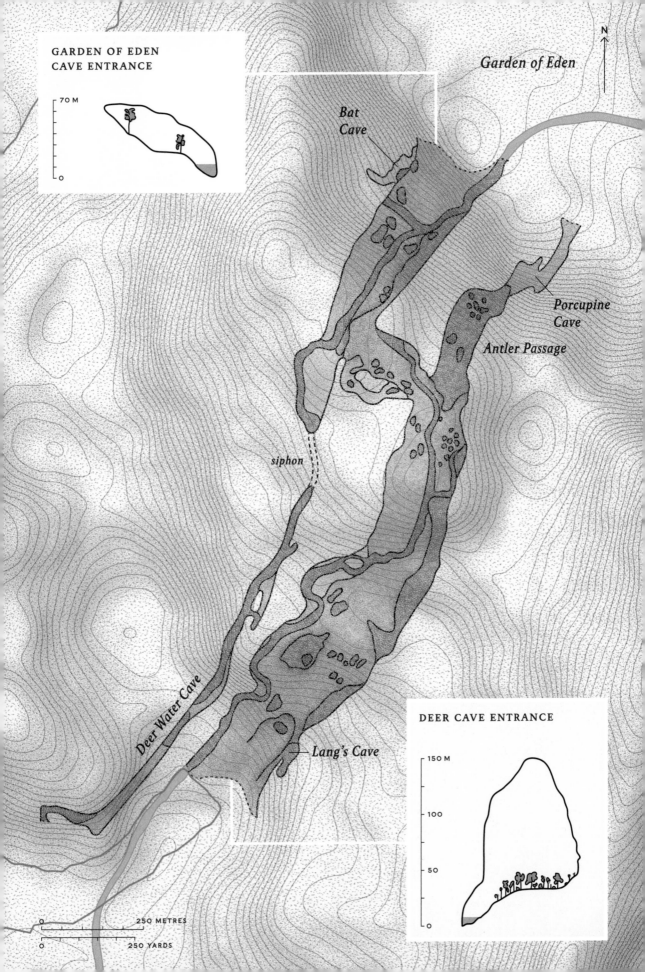

N

GARDEN OF EDEN
CAVE ENTRANCE

70 M

0

Garden of Eden

Bat
Cave

*Porcupine
Cave*

Antler Passage

siphon

Deer Water Cave

Lang's Cave

DEER CAVE ENTRANCE

150 M

100

50

0

0   250 METRES

0   250 YARDS

named for the animals that had been observed coming there to take shelter. With the indigenous Berawan claiming to have no knowledge of anyone going all the way through the cave before, the Hanbury-Tenisons could quite legitimately claim to be the first people to ever set foot in the world they found at the other end.

Deer Cave has since become renowned for far more than just its size, with the diversity of its fauna also highly remarkable. As first observed during the expedition, the departure and return of a huge colony of bats every evening — estimated to number over a million — has become a famous characteristic of Gunung Mulu National Park. Small critters scutter around the caves, such as the eye-catching phosphorescent centipedes. And the avian life that inhabits the cave is so numerous that an entire industry has materialised based around small groups of local harvesters climbing up rickety ladders to carve away chunks of the nests of cave swiftlets. This is the central ingredient in the highly popular bird's nest soup, considered something of a Chinese delicacy. As such the raw material is valued to be worth its weight in silver (and poses a threat to the survival of many of the birds in question).

Deer Cave is certainly not the only massive geological cavity in this part of the world. Indeed, the landscape is riddled with large, unexplored caves and passageways, from Green Cave, whose entrance was found so stuffed full of vegetation that the sparse light inside is endowed with a green hue, to Sarawak Chamber, the largest known cave chamber in the world (over eleven Great Pyramids of Giza would fit inside it). Indeed, the subterranean network of caverns and cavities in Mulu is believed to cover more than 600 kilometres (370 miles), due to the geological processes that led to the dramatic uplift the region experienced between 2 and 5 million years ago, hoisting Borneo out of the ocean and twisting it in a way that resulted in the medley of caves still being discovered today.

The first known people to pass all the way through Deer Cave described finding a lost
← *world at the other end.*

Pando

AN ANCIENT FOREST OF CLONED TREES THAT IS CONNECTED UNDERGROUND

UTAH, USA

N 38° 31' 35"
W 111° 44' 57"

Fields have eyes, and woods have ears, according to both Geoffrey Chaucer and centuries of old wives. And if trees do possess instruments with which to hear, could it be that they also possess instruments with which to speak? Indeed, there is a growing body of evidence — actual hard science — to support the idea of talking trees, and listening ones too. It seems that forests contain much more chitchat and whispers through the leaves than hard-nosed scientists have previously believed.

Perhaps this doesn't happen in the way we might envisage from *The Lord of the Rings* or *Pocahontas*, but certainly it is increasingly accepted that many trees are communicating, trading and transferring sugars and nutrients through an interconnecting root network below ground. The roots themselves aren't connected, incidentally; instead thin, stringy fungi called mycorrhizae act as middlemen, joining roots from different trees together and passing chemical messages between them. In this way, healthy trees can support and even keep alive their sick or damaged neighbours, thereby sustaining the health of the entire forest. It is even believed that when one tree is cut or its leaves or bark eaten, it can notify those trees nearby — in what we might dramatically imagine as a silent subterranean scream — so that they can prepare to fend off an attack themselves, perhaps by increasing the toxicity of their leaves, for example. This 'wood wide web', as some are

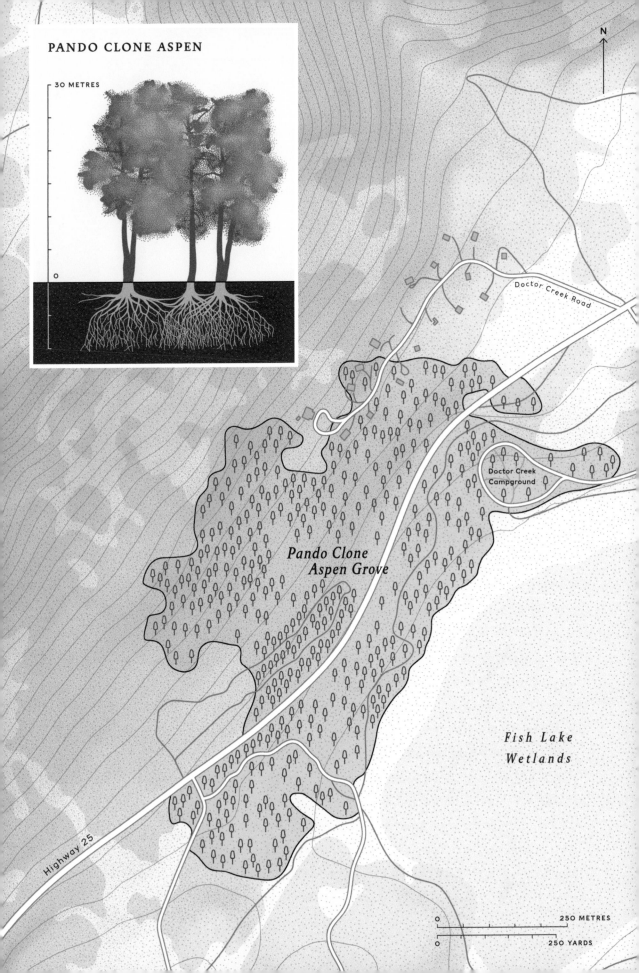

PANDO CLONE ASPEN

30 METRES

0

Doctor Creek Road

Doctor Creek
Campground

*Pando Clone
Aspen Grove*

*Fish Lake
Wetlands*

Highway 25

N

250 METRES

250 YARDS

romantically calling it, is just one of the previously unknown mysteries of dendrological communication that are slowly being uncovered.

There is one part of the world where this subsurface passing of notes takes on a whole new level of significance. In Fishlake National Forest, near the town of Richfield, southern Utah, United States, lives the world's largest single organism. It is a quaking aspen forest, so called for the 'quaking' sound the trees produce when filtering passing wind through their leaves. Spanning 43 hectares (106 acres), the forest goes by the collective name 'Pando'. It is estimated that these roughly 47,000 trees — each around 30 metres (100 feet) tall, and famously vividly bright golden and scarlet in their autumnal colours — are not just connected to each other below the earth's surface, in the aforementioned manner of most trees, but are in fact a single organism, perhaps the largest in the world. The trees are clones, clones that happen to have multiplied over tens of thousands of years.

How can this be possible? What's going on down there that enables this bizarre process to occur? The answer again lies underground. Instead of the usual system whereby a single plant sexually reproduces by fusing its DNA with that of a neighbouring tree through the clever dispersal of pollen, Pando's trees share a common root system below the surface, as is often typical of aspens. These roots are capable of taking on a somewhat investigative role, exploring their surrounding environment, and then asexually 'reproducing' — or at least producing what might appear to be an entirely new tree — next door. This new tree, still connected to the parent below the surface, can grow to maturity and then repeat the practice.

Multiply this bizarre process several thousand times over millennia, and you get a sprawling aspen forest, all linked together by a chain of roots below ground, with both the ability and willpower to colonise the landscape that surrounds it. Expansion is further helped by the quaking aspen's unique quality of possessing chlorophyll — vital for photosynthesis to occur — not just in its leaves, like most trees, but also in its bark. Therefore, even in

↑
Thanks to a vast shared root system, this quaking aspen forest is the world's largest organism.

winter, when its leaves lie dead and useless on the forest floor, the aspen is capable of continuing its photosynthesising, and subsequent growth.

Yet Pando is threatened. The only way this organism can survive is to continually produce new shoots and successfully turn them into new trees, to offset those lost over the years (each 'tree' can live for around a century, perhaps a century and a half). Unfortunately, this process has recently become seriously inhibited: satellite imagery shows a gradually degrading forest over the past thirty to forty years.

The main culprits appear to be the mule deer (and sometimes cattle) that have made this forest their home since the decline in native predators such as grey wolves, primarily eliminated by wary humans, allowed the population of these grazers to explode. Their insatiable appetite for delicious young shoots to feed upon has prevented Pando from producing new trees, causing the forest as a whole to become aged and increasingly sick. A growing human presence in the form of power lines, camp grounds, hiking trails and cabins only adds to the problem.

The consequence is that we are witnessing the demise of an organism whose very existence, as a creature rooted into the earth — whose fingers rise towards the air to appear as a standard forest, without exposing its true heart below the soil — makes us question what we still don't know that we don't know about the natural world.

Dark Star Cave

POSSIBLY THE WORLD'S DEEPEST
HIGH-ALTITUDE CAVE

UZBEKISTAN

N 38° 23' 47"
E 67° 17' 13"

In 1984, a Soviet team from the Sverdlovsk Speleological Club were exploring a remote spot in the Boysuntov (also known as Baysun-Tau) Mountains. Close to Uzbekistan's Afghan border, it's nearly 400 kilometres (240 miles) from the capital Tashkent as the metaphorical crow flies, and perhaps twice that in real-world land travel.

At a sheer limestone cliff named Xo'ja Gurgur Ota that loomed 365 metres (1,200 feet) high above their heads — the side of a 35-kilometre-long (22-mile) plateau that runs across the rugged landscape — they began following their curiosity and delving into a mysterious hole high above. It was to be the first known entrance into a world of phenomenal subterrestrial grandeur — perhaps the world's deepest high-altitude cave.

But this team did little further to explore their discovery, and so it was left to a British crew named Aspex '90 to turn up six years later and see what geological treasures lay inside this cliff face. With a quirky sense of humour, it was these explorers who named the entire underworld after an obscure, cult science-fiction movie of the 1970s named *Dark Star*, about the crew of a ship lost in space. Perhaps it tells us something about the emotions and sense of disconnection they felt during their days and nights inside the cave system. Ultimately, they were restricted in how far into the earth they could burrow — so deep was the interior that they

←
*Scaling steep,
high-altitude cliffs
is a prerequisite
for descending into
Dark Star Cave.*

found themselves ill-equipped for the challenge at hand. The next year, a follow-up crew, Aspex '91, tried to continue their progress, but were prevented from doing so by melting ice. The imminent collapse of the Soviet Union then delayed any further investigation into this region of the world for many years.

Decades later, Dark Star has become a hotspot for modern cave explorers, with a total of eight expeditions to date mapping 18 kilometres (11 miles) of tunnels and remarkable geological formations. There are now seven known entrances within the cliff (Orenburgskiy is the highest of these, sitting at 3,590 metres (11,778 feet) above sea level at the very top of the cave, just above

▲ XO'JA GURGUR OTA

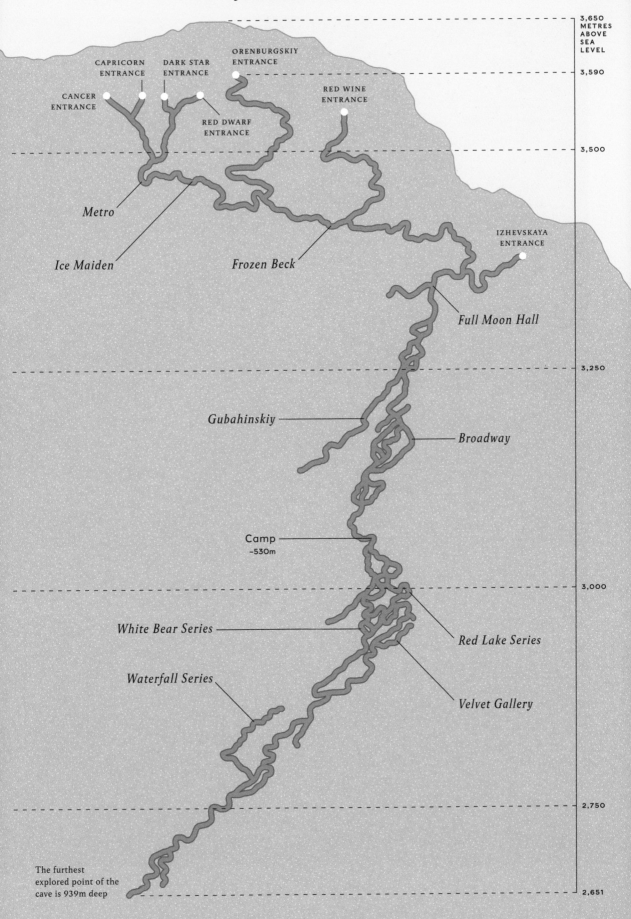

3,650
METRES
ABOVE
SEA
LEVEL

ORENBURGSKIY
ENTRANCE

3,590

CAPRICORN
ENTRANCE

DARK STAR
ENTRANCE

RED WINE
ENTRANCE

CANCER
ENTRANCE

RED DWARF
ENTRANCE

3,500

Metro

Ice Maiden

Frozen Beck

IZHEVSKAYA
ENTRANCE

Full Moon Hall

3,250

Gubahinskiy

Broadway

Camp
–530m

3,000

White Bear Series

Red Lake Series

Waterfall Series

Velvet Gallery

2,750

The furthest
explored point of the
cave is 939m deep

2,651

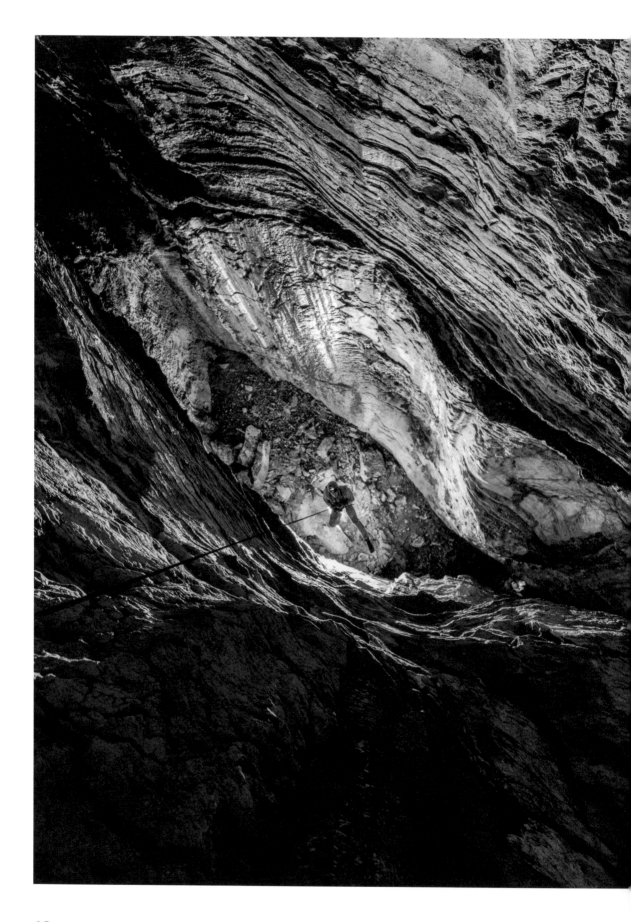

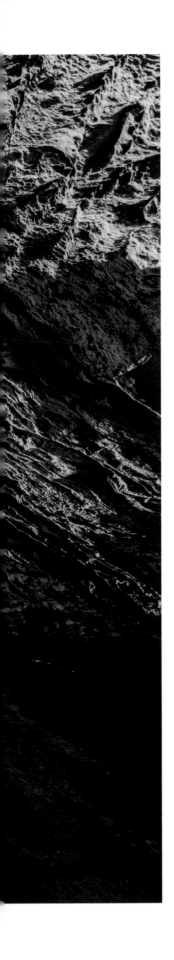

the original entrance). None is accessible without professional rock-climbing experience and proper equipment. To enter the cave, it is necessary to scale Xo'ja Gurgur Ota to at least Izhevskaya, the lowest of the cave's known entrances, at 137 metres (450 feet) vertically in the air, before rappelling down once on the inside. Even aiming for this lowest opening requires several days to hook up the necessary ropes for a full-scale expedition.

Perhaps surprisingly, given the sweltering heat to be found underneath the baking sun across the Boysuntov, the air that emerges from the cave entrances is bitterly cold. Indeed, these freezing temperatures define much of the interior of Dark Star. Despite being thousands of metres above sea level, the cave experiences extremes perhaps more associated with subterranean underworlds, from walls covered in icy crystals to large lakes full of solid ice. Huge waterfalls can be found almost frozen in time, their streams turned to thick icicles.

Just after the final waterfall, at an estimated top-to-bottom depth of 939 metres (3,080 feet) — still 2,651 metres (8,698 feet) above sea level — a tiny crack in the rocks, mere inches wide, appears to be the only way to progress forwards. On the other side, a new tunnel, heading into the unknown. Perhaps this is the fabled connection with neighbouring Festivalnaya, another long cave network? If so, the size of their combined cave system would begin edging into truly record-breaking territory. Unfortunately, this is an endeavour left for future explorers, for an expedition as yet unwritten.

No one knows how much further this cave network extends, or whether it might even connect with
← *a neighbour.*

Palaeoburrows

MYSTERIOUS TUNNELS BELIEVED TO HAVE BEEN DUG BY EXTINCT MEGA SLOTHS

BRAZIL

S 29° 45' 27"
W 53° 17' 21"

In late 2008, Heinrich Frank, a geologist at the Federal University of Rio Grande do Sul, Brazil, was being driven along a highway near the town of Novo Hamburgo when he noticed a strange hole within an excavation by the side of a road. Something about the unusually circular shape caught his attention. On this occasion, he had no time to ask his driver to stop, so his geological inquisitiveness would have to wait. But a few weeks later he passed by again while driving with his family and was able to stop the car to take a closer look. It was a tunnel, he noticed with growing curiosity, in the middle of dense clay sandstone. What could possibly have created such a strange feature? wondered Frank, as he scrambled deeper into the earth.

Lest we forget, the vast majority of all species that have ever existed on earth are now extinct, some more famously than others. Mammoths and sabre-toothed tigers might be household names, but they were joined in the Pleistocene era by the equally enigmatic but lesser-known mastodons, cave lions and giant beavers. In many cases, disintegrated fossils and skeletons are all we have left to decipher everything we know about certain species, their appearance, habitats and behaviour, thousands of years since they departed this world.

Other species leave such clear evidence of their existence, it can take a while for humans to even recognise their contribution.

N

COLOMBIA

VENEZUELA

GUYANA

SURINAME

FRENCH
GUIANA

*NORTH
ATLANTIC
OCEAN*

PERU

B R A Z I L

BOLIVIA

PARAGUAY

CHILE

*SOUTH
ATLANTIC
OCEAN*

URUGUAY

ARGENTINA

● *Palaeoburrow*

0   1,000 KILOMETRES

0   500 MILES

One such case is demonstrated by Frank's trips into the Brazilian undergrowth. Millions must have passed by this spot on the highway for years, but he was the first with the inclination (and the expertise) to stop and explore further, to wonder how this hole was actually created. The interior of the tunnel was somewhat elliptical, around a metre (3 feet) in diameter, a shape that Frank knew would never have been formed by water. Most tellingly, there were gashes on the surrounding walls digging deep into the rock, like huge claw marks, features unlike anything he had seen before.

Unlike evidence of geological processes such as the rushing of water and the passing of time, or violent tectonic activity, these

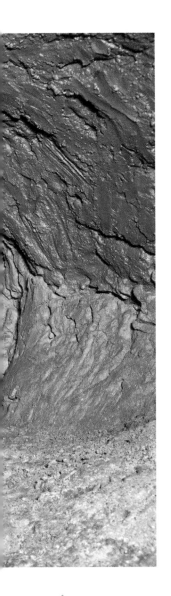

are tell-tale signs of an animal at work, one that ceased to exist around 10,000 years ago. Giant ground sloths were the likely culprits — creatures believed to have once been widespread across South America, officially named *Megaichnus* igen. nov. Like the modern-day giant armadillo, these animals are thought to have been expert tunnel diggers and, given their substantial bulk — perhaps comparable to modern-day elephants — they required an extensively large tunnel to shelter in, up to 1.5 metres (5 feet) in diameter.

But this tunnel wasn't unique. Having become aware of the existence of palaeoburrows, Frank embarked upon a mission to uncover more such formations. Over the subsequent decade he recorded more than 1,500 palaeoburrows in Rio Grande do Sul alone, while many more have been spotted in other Brazilian states, such as neighbouring Santa Catarina, along with a handful in northern Paraná. Their spread gives some evidence regarding the historic range of the giant sloth, with their presence appearing to be almost entirely limited to southern Brazil, with just one recorded in Uruguay and Paraguay, and none outside the Americas at all. From small, easily missed clues, such large revelations can emerge.

↑

Thousands of palaeoburrows such as this have been discovered across southern Brazil, scratched from the earth thousands of years ago.

Giant ground sloths were prehistoric animals who left their mark on the contemporary landscape.

→

Ancient History

Chauvet-Pont d'Arc

HOME TO SOME OF THE EARLIEST AND BEST-PRESERVED FIGURATIVE DRAWINGS IN THE WORLD

FRANCE

N 44° 23′ 21″
E 04° 24′ 57″

Many animals on the walls have been extinct in Europe for tens of thousands of years, further indicating the advanced
← *age of the drawings.*

When ancient humans talk to us, subterranea is their medium of choice. Underground spaces have become so synonymous with *Homo erectus*, *Homo neanderthalensis* (more commonly known as Neanderthals) and other early hominids that they are used as a shorthand way to describe them: cavemen.

And so to the netherworld we go to discover more about our primeval cousins, those ancestors whose genetic codes we carry around in contemporary twenty-first-century life. Much of what we know about the lives of those from whom we descended has been obtained by studying the bones, tools, animal remains and, of course, artworks left by these prehistoric peoples in caves, the places they most commonly inhabited for safety and shelter. Many of the most significant discoveries about early humans occurred in such underground spaces, from Liang Bua Cave in Flores, Indonesia, where *Homo floresiensis*, the famous 'hobbit' species, was discovered in 2003, to Denisova Cave in Russia, where the as-yet-unclassified 'Denisovans' were found in 2008, to the 2012 discovery of the 'Red Deer Cave people' of southwest China.

It is with this in mind that we transport ourselves to a limestone plateau in southeast France. It's late December 1994, and Jean-Marie Chauvet, a local park ranger with the Ministry of Culture, is with friends Eliette Brunel and Christian Hillaire, hunting for prehistoric artefacts in a gorge above the Pont d'Arc,

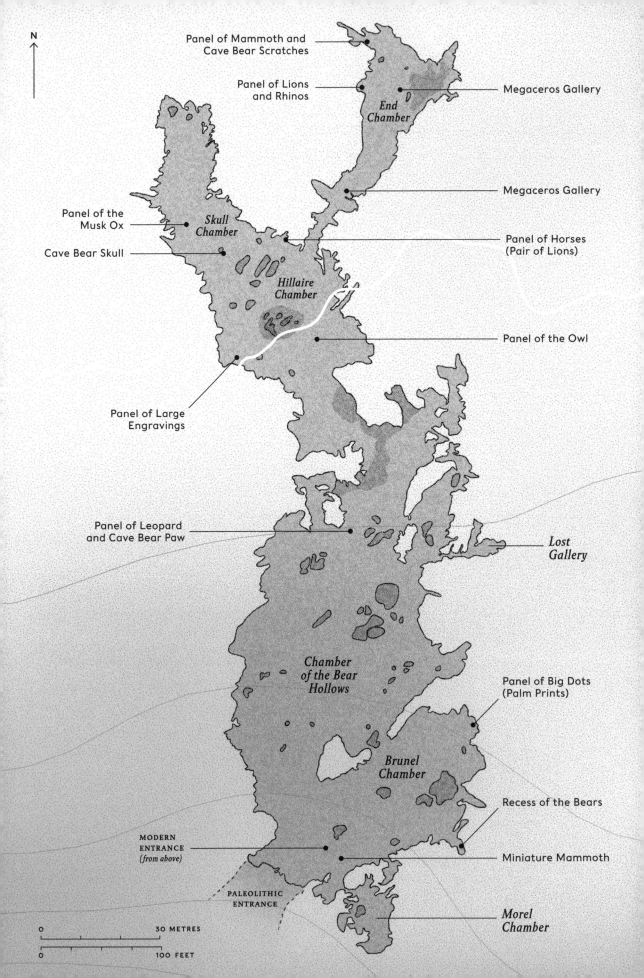

N

Panel of Mammoth and
Cave Bear Scratches

Panel of Lions
and Rhinos

Megaceros Gallery

*End
Chamber*

Megaceros Gallery

Panel of the
Musk Ox

*Skull
Chamber*

Cave Bear Skull

Panel of Horses
(Pair of Lions)

*Hillaire
Chamber*

Panel of the Owl

Panel of Large
Engravings

Panel of Leopard
and Cave Bear Paw

*Lost
Gallery*

*Chamber
of the Bear
Hollows*

Panel of Big Dots
(Palm Prints)

*Brunel
Chamber*

Recess of the Bears

MODERN
ENTRANCE
(from above)

Miniature Mammoth

PALEOLITHIC
ENTRANCE

*Morel
Chamber*

0   30 METRES

0   100 FEET

a natural bridge that spans the Ardèche River. Stumbling upon a narrow gap in an apparent rockfall, Chauvet takes the bold decision to squeeze his body through the small hole. Inching his way down the mysterious passageway, he drops into a previously undiscovered cavity beneath the earth, quickly followed by his two friends. The floor is covered in animal bones, broken stalagmites and other scattered debris. Suddenly Brunel calls out, 'They have been here!' after spotting two small red parallel markings on a rock. With a shock, the three look up to see the walls surrounding them covered with hundreds of vivid artworks, mainly paintings and engravings depicting enigmatic animals.

Given their professional experience, even in this moment the three might have been aware that they had just become the first people in thousands of years to set eyes upon the oldest known cave drawings in the world. The drawings — a thousand of which have so far been recorded for posterity — have been radioactively dated to at least 30,000 years ago, and are potentially even older. While these results remain somewhat controversial, subsequent dating of the bear bones also found in the Chauvet-Pont d'Arc cave (as it is now ceremonially known) produced an estimated age of between 29,000 and 37,000 years old, supporting the initial analysis.

Given the sophisticated styles and techniques adopted in the drawings — including the use of bulges in the cave wall and different-sized animals to create perspective, and a rich usage of colours — the discovery has, in the words of the French Ministry of Culture, 'overturned the accepted notion about the first appearance of art and its development'. Fascinatingly, the depicted animals — presumably drawn by the Aurignacian people who once lived here — are significantly different from those typically seen on the interior of cave walls. While there are the standard horses and bison seen in other parts of Europe, at least half the sketches show animals perhaps not commonly associated with the continent, including mammoths, lions, hyenas and even rhinos. Brought back to life by art, it's a glimpse into a lost world, a time when these wild and dangerous animals would have freely wandered the European landscape, before improved hunting techniques by our forefathers led to their extinction from the continent.

Sadly, experience shows that allowing mass tourism into these unique and delicate spaces is among the fastest ways to eliminate everything that's special about them. Lascaux and Altamira — fellow specially recognised cave regions full of prehistoric artworks in the Dordogne, southern France, and Cantabria, northern Spain, respectively — are both now closed to the public, but not before decades of carbon dioxide, heat and humidity brought in by tourists had damaged their precious masterpieces for ever. The rockfall Chauvet once burrowed through may have protected Pont d'Arc's drawings for 20,000 years or more, but exposure to the humidity and air composition of the outside environment, and the steady arrival of more and more experts and researchers, threaten to irrecoverably damage them too. While it was decided that the discovery was too rare and invaluable to ever open Chauvet-Pont d'Arc to the general public, the organisation ERGC (Grand Projet Espace de Restitution de la Grotte Chauvet), in collaboration with Google, has instead created a fully accurate recreation of the cave, complete with detailed drawings, for visitors to enter and experience virtually instead.

The rich styles appearing on the walls at Chauvet revolutionised our understanding of the development of various artistic techniques.
→

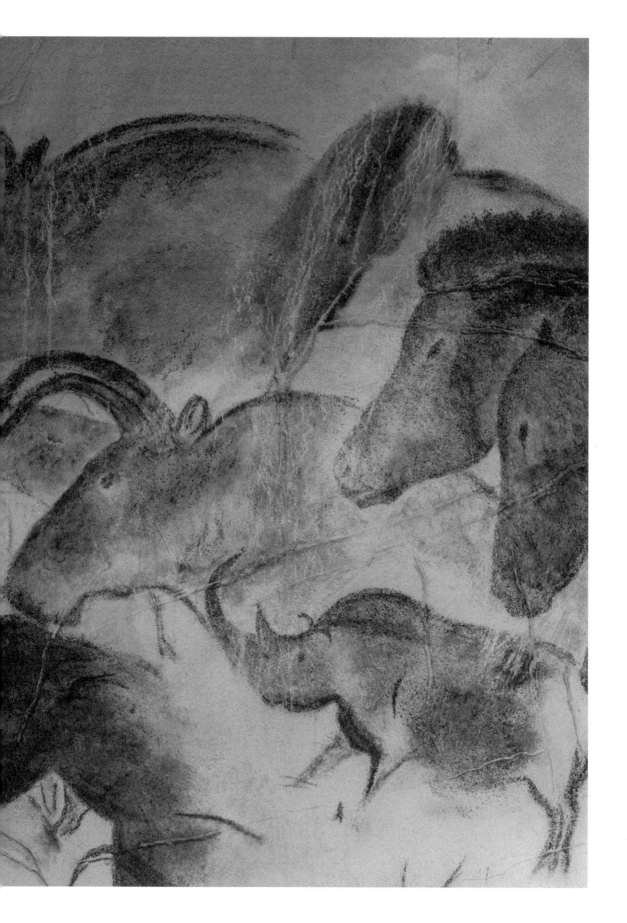

Derinkuyu

AN IMMENSE NETWORK OF UNDERGROUND CITIES, ONCE HOME TO 20,000 PEOPLE

TURKEY

N 38° 22' 31"

E 34° 44' 01"

You can run, but you can't hide, according to bad guys throughout movie history. The ancient residents of Cappadocia, in modern Turkey, might beg to differ. When conflict came charging over the horizon, they would take to ground, scurrying away like rabbits retreating into their warrens. Once the invading armies arrived on the scene, they'd be confronted with nothing but a deserted ghost town, comprising a series of confusing and seemingly impenetrable caves. It was as though thousands of people had vanished into dust.

The landscape of Cappadocia has a bewildering quality, where the dividing line between natural and human creation blurs into obscurity. Here in Central Anatolia, north of the Taurus Mountains, lies seemingly alien scenery. Created through millions of years of volcano eruptions coating an old lake with layer upon layer of ash, the land eventually cooled into a soft, porous rock hundreds of feet deep known as 'tuff', covered by lava that cooled into a hard topping of basalt. So-called 'fairy chimneys', pillars of tuff left behind when the land around them eroded away to the point of collapse, now protrude from the landscape like nails jutting haphazardly from an old piece of wood. Some are as tall as 40 metres (130 feet), each with its own little mushroom-style basalt hat on top.

Such uniquely remarkable geology is part of why Cappadocia is now part of the Göreme National Park, a UNESCO World Heritage site since 1985. But that's just the surface view. Below ground, things get far more bizarre. Cappadocia contains an astounding 250

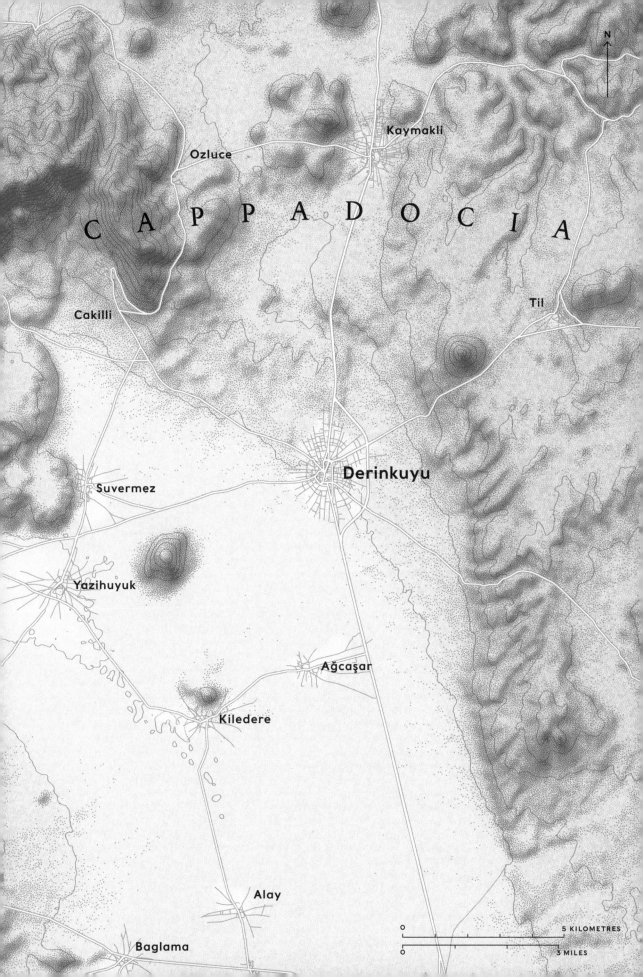

subterranean towns and cities and more, carved by human hands from the malleable tuff terrain that the volcanoes left behind.

The largest of these cities is Derinkuyu, supposedly not discovered until 1965, when one above-ground resident knocked down a wall in his home and found an entrance to an immense maze below (Derinkuyu literally translates as 'deep well'). Once excavated, Derinkuyu was revealed to comprise dense tunnels like ant farms stretching as deep as 85 metres (280 feet) below the surface — around the height of Big Ben or the Statue of Liberty — and containing more than eighteen levels. These included vast storage spaces, schools, churches, and even fully equipped underground wineries and breweries fed by water from the deep subterranean river running below.

Large vertical shafts ran between levels, some functioning as wells to enable residents to access this fresh water, while others gave a method of easy communication between floors. To overcome the eternal darkness, linseed oil lamps would have illuminated the warrens and provided mild warmth. It is estimated that Derinkuyu was once home to in excess of 20,000 people.

And it wasn't just humans who made their home underground. Special chambers were dug out to enable livestock to join their owners in these tunnels, including large stables for horses. By adding lime to cow livers, residents devised a method for speeding up the decomposition of organic waste, enabling them to dispose of the immense quantities of human and animal excrement that would otherwise overrun the tunnels. Furthermore, in a macabre twist, the tunnels even contained temporary underground tombs capable of storing the bodies of the deceased, until it was safe to return to the surface and give them a proper burial.

Living in a region of the world prone to waves of attacks from foreign invaders passing between Europe and the Middle East (and vice versa), like a small boat caught in rough waters, the arrival of undesired foes was a regular concern for the residents of Cappadocia. Perhaps the founders of Derinkuyu and neighbouring cities were the Hittite Empire, premodern people who dominated the region between 3,000 and 4,000 years ago. It is

A cross-section of the Derinkuyu tunnels, stretching as deep as 85 metres below the surface.

←

Natural holes in the Anatolian landscape hint at the remarkable settlement to be found inside.

believed they dug out the first, somewhat rustic upper levels, as crude defences against aspiring invaders such as the Phrygians.

In time, the Hittites faded from history, leaving Cappadocia to be superseded by the Assyrians, the Persians, and so on, each civilisation digging deeper than the last. The Byzantines took control of the region around the sixth century, vastly expanded Derinkuyu's tunnels, digging ever deeper, and creating the smooth, more professionally rectangular lower levels. The Byzantines truly dug in their heels to ensure their continued presence in the face of centuries of Arab invasions, devising a series of medieval tricks, traps and barricades to dissuade potential assailants.

First, the earth ovens in Derinkuyu's large communal kitchen had chimneys that were deliberately skewed and channelled to release fumes above ground over a mile away, so as not to give away the city's true location. If invaders did find the entrances, they were forced through narrow, low tunnels, demanding they hunch over, maybe even crawl. Tunnels were fitted with giant wheels that could be rolled across entrances to prevent entry or

Ingenious booby traps were installed in the entrances to the underground city, to ensure complete security in the event of an invasion.

→

exit, trapping invaders inside. Small holes in the wheels enabled defending forces to spear their opponents from the safety of the other side, while pipes installed above their heads would shower them with scalding-hot oil. A now-collapsed tunnel even stretched for 10 kilometres (6 miles) to connect Derinkuyu with the neighbouring underground city of Kaymakli, home to an estimated 15,000 residents, providing a potential escape route. All these cleverly devised traps and back-up strategies made it possible for Derinkuyu's thousands of residents to take refuge in the safety of the subterranean world for weeks, or even months, if necessary.

In 2013, construction workers undertaking demolitions in the nearby region of Nevşehir discovered unknown tunnels leading deep underground, where a brand-new subterranean city was uncovered. With large churches, kitchens, wineries, linseed oil-pressing facilities and more, it has been speculated that the new city could rival or even exceed the scale of Derinkuyu. University geophysicists have surveyed the area and calculated it to plunge to an estimated 113 metres (371 feet), far deeper than Derinkuyu. The mysteries of Cappadocia continue to reveal their secrets.

Mausoleum of the First Qin Emperor

AN UNOPENED TOMB, GUARDED BY THOUSANDS OF TERRACOTTA SOLDIERS

CHINA

N 34° 22' 53"
E 109° 15' 14"

Crack. After navigating a series of ingenious Indiana Jones-style booby traps, their goal is finally in sight. The centre of the mysterious tomb is just inside this looming crypt, sealed for over 2,000 years ... until now. The intrepid excavator team have a crude but direct path to their target, by breaking open the tomb by force. And yet as they cleave the mausoleum in two, members of the team begin coughing, having difficulty breathing. One turns to the other, her eyes red and swollen. Shaking, she clutches her chest, feeling desperately for the pains that suddenly plague her. One by one, the team begin collapsing, each coughing up blood. After more than two millennia, the Mausoleum of the First Qin Emperor continues to claim victims.

The above events have not occurred; they are pure fiction, complete with a smattering of Hollywood-style drama. But they are not so far removed from real-life speculation about what might occur in the event of opening the Mausoleum of Qin Shi Huang, the final resting place of the first emperor of China's Qin dynasty. In a truly prolific reign, Qin united rival tribes across a huge area to become one, unified Chinese nation, and then began construction of a Great Wall to the north. The English name for the country is even derived from his own.

Some emperors demand their own private army, to maintain their security and luxurious prosperity for life. Others continue

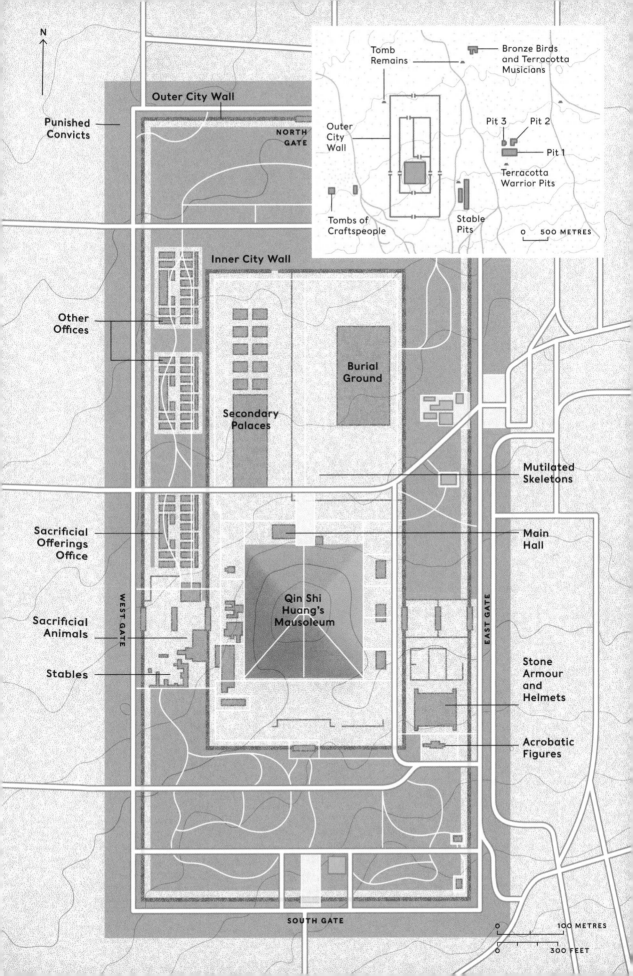

N

Punished
Convicts

Outer City Wall

NORTH
GATE

Tomb
Remains

Bronze Birds
and Terracotta
Musicians

Outer
City
Wall

Pit 3   Pit 2

Pit 1

Terracotta
Warrior Pits

Tombs of
Craftspeople

Stable
Pits

0   500 METRES

Inner City Wall

Other
Offices

Secondary
Palaces

Burial
Ground

Mutilated
Skeletons

Sacrificial
Offerings
Office

Main
Hall

Qin Shi
Huang's
Mausoleum

WEST GATE

EAST GATE

Sacrificial
Animals

Stone
Armour
and
Helmets

Stables

Acrobatic
Figures

SOUTH GATE

0   100 METRES

0   300 FEET

such expectations even after their departure from this world. Such was the case with Qin Shi Huang. Upon his death in 210 BCE, there began the construction of clay soldiers, thousands of them, to protect the great emperor in the afterlife. Since first uncovered by rural farmers in 1974, over 2,000 figures have been dug from the earth, with up to another 6,000 believed to still be buried. This entourage of bodyguards (although also including civil servants such as musicians and acrobats) is so vast, impressive and strangely well preserved that the Terracotta Army, as it is now commonly known, has become even more famous than the great emperor whom they were assigned to guard over. Since being unearthed, many have gone on tour, with the first overseas exhibition being held in Melbourne, Australia, in 1982, and they have subsequently been displayed in museums around the world, from Sydney to Santiago, New York to New Delhi, Toronto to Turin.

But while the army basks in its international celebrity status, the man behind its existence remains something of a mystery — or, at least, his burial place does. Critically, the tomb has never been opened. As it is considered perhaps the most significant archaeological discovery since Tutankhamen in Egypt in the 1930s, authorities have reserved extreme caution regarding its treatment, keen to learn lessons from this and other somewhat botched excavations from the past. Despite the passing of over four decades, they continue to bide their time, patiently awaiting the arrival of technology that will allow them to learn significant details about both the tomb and the underground city surrounding it, before attempting to physically enter.

The scale of the unknown is staggering. The tomb itself is 76 metres (250 feet) high, and spans an area of around 172,500 square metres — over 42 acres, roughly two Buckingham Palaces. Partially completed 3D images obtained by volumetric scanning show a large courtyard with at least eighteen houses, loomed over by one central building: presumably the final resting place of Qin himself. While such a tomb may seem elaborately large, it's dwarfed by the size of the site that surrounds it. Overall, the entire excavation covers an area of around 56 square kilometres (over

Qin Shi Huang was buried in 210 BCE with an estimated 8,000 terracotta soldiers and servants to protect him in the afterlife.

→

← *The ornate and well-preserved details of the thousands of figures in the Terracotta Army have made them one of the world's most famous examples of funerary art.*

21 square miles) — twice the size of the territory of Macau. All constructed entirely for one deceased emperor.

Chief among archaeological concerns is the possibility that the Terracotta Army represents a serious warning about actual, lethal dangers that lie ahead for anyone attempting to enter the tomb. More scary than a clay soldier, genuine traps may have been set that are potentially still active. Studies by local researchers

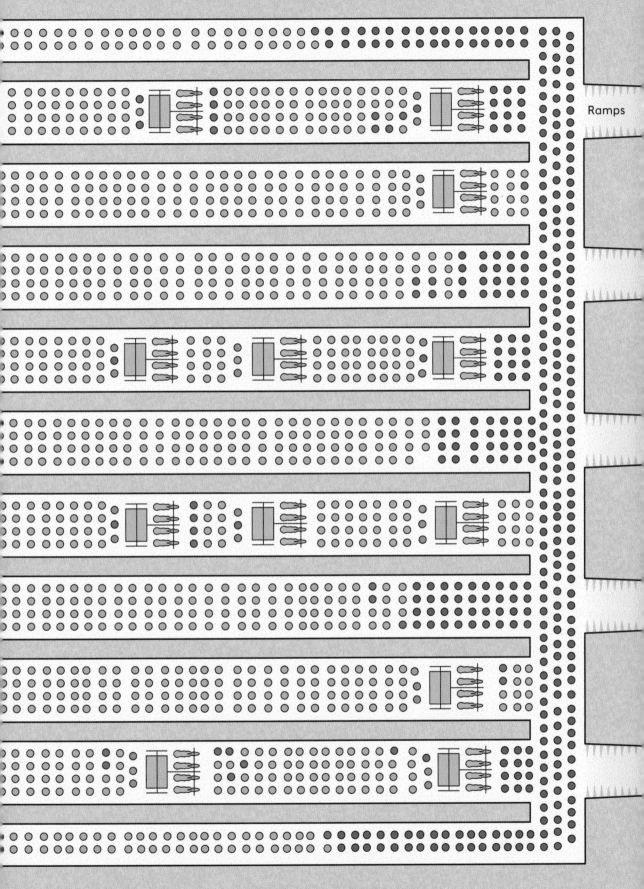

TERRACOTTA ARMY PIT 1, EASTERN SECTION

Ramps

○ Armoured Soldier   ● Unarmoured Soldier   ● Officer   ○ Charioteer   ○ General

have confirmed extremely high mercury levels detected around the vicinity of the tomb, leading to speculation that deadly mercury poisoning might await aspiring tomb raiders. Hence the desire to rely on — among other technology — tiny robots capable of entering the tomb first for initial analysis, to see how outlandish (or not) these suggestions might be.

Such a situation brings to mind contemporary debates about how to tell future generations not to interfere with certain burial spots. Assuming that current languages and symbols eventually become obsolete, how could modern society warn a civilisation 2,000 years or more from now not to open a sealed vault containing, for example, toxic nuclear waste, contaminated mine pollutants, or perhaps the remnant pathogens of a destructive plague? Suggestions including integrating symbolically intimidating architectural designs, such as brutal spikes, or perhaps embedding the threat into community culture by crafting and passing on discouraging songs and folklore tales, have been proposed as possible strategies to warn our descendants not to look into certain buried hazards. And yet, as the case of the Mausoleum of China's First Qin Emperor shows, perhaps the persistent curiosity of humanity means it will always be an uphill task to ask people to simply leave such buried places alone.

Herculaneum

ITALY

N 40° 48' 21"
E 14° 20' 51"

A monastery, situated in Resina, a coastal town in the Bay of Naples, in the year 1709. A workman sweats, digging a hole in the courtyard to turn into a well for the monks to refresh themselves. Surprisingly, he discovers pieces of marble. He knows this material is in demand in the construction of a nearby luxury villa, so he alerts the Austrian prince on whose orders the villa is being constructed. The prince, enthused about this discovery, orders the well to be expanded, and the valuable materials inside removed. Thus begins, in this amateurishly unintentional fashion, the excavations that uncovered the buried remnants of the town of Herculaneum, lost from history for over one and a half thousand years.

The reason for Herculaneum's dramatic disappearance is well known, although primarily for the impact on a neighbouring settlement. On 24 August in the year 79 AD, as the sun rose on a bright Italian morning, Mount Vesuvius erupted, blasting a thick cloud of ash, rocks and other debris into the air. 'Sometimes it looked white, sometimes blotched and dirty, according to the amount of soil and ashes it carried with it,' described one witness, the famously eloquent Roman writer Pliny the Younger. His recollection of the cloud noted its resemblance to a pine tree, 'for it rose to a great height on a sort of trunk and then split off into branches'.

Ash and pumice rained down upon the Roman city of Pompeii. Those who couldn't flee were forced to share the fate of the once-wealthy town, caught in the dropzone of the falling volcanic debris that covered much of the urban landscape, crushing buildings and killing thousands. Many people were frozen in time, caught in the terrified crouches or huddled positions they adopted in their final moments. Pompeii has since become one of the most iconic sites of archaeological anthropology in the world.

The small coastal town of Herculaneum experienced a similarly apocalyptic fate. As Pompeii slowly disappeared underneath layer upon layer of ash, Herculaneum, something of an opulent holiday resort, and at that moment in a pomp festival mood, lay in the path of the immensely powerful pyroclastic flow that burst forth from the volcano's summit. Travelling at a likely 100 kph (60 mph) or more, with temperatures calculated to have reached at least 520°C (968°F), the current of hot gas and volcanic matter would have utterly overwhelmed the small Greek-built settlement. In only a matter of minutes, what was once a thriving party town vanished under the cloud of devastation that Vesuvius wreaked upon the Mediterranean coastline on that warm summer's day.

And so Herculaneum was virtually wiped from existence, from even the collective imagination of the descendants of the citizens who survived these traumatic events. New settlers moved in, building upon land in the shadow of Mount Vesuvius that appeared to be left neglected. A new town, Resina, unknowingly sprawled atop the debris that had once buried Herculaneum.

The excavations of the early eighteenth century changed everything. It would have been a shock to local residents to find out that beneath their streets lay the dormant ruins of a once-great town, destroyed by the power of the volcano that loomed above them. But the knowledge ignited a desire to pull back the ground still further to see the ancient world buried beneath the residents' feet.

For over 150 years, sporadic excavation took place to unearth the secrets buried in Herculaneum. But it took until 1927 before state funding became available to begin a proper dig to get beneath Resina (renamed Ercolano in the middle of the twentieth century), into the turf earth that the pyroclastic flow had laid down.

Modern
Buildings

BORDER OF EXCAVATION SITE

Decumanus Maximus

N

A

INSULA
VI

B

C

INSULA
V

D

E

INSULA
ORIENTALIS
II

F

G

Cardo III

Cardo IV

Cardo V

Decumanus Inferior

I

INSULA
III

J

H

K

INSULA
IV

L

INSULA
ORIENTALIS
I

M

QUARTIERE
SUBURBANA

Original Coastline 79 AD

0   25 METRES

0   75 FEET

A. Shrine to Augustus

B. Forum Baths

C. House of the Beautiful
Courtyard

D. House of the Charred
Furniture

E. Upper Hall

F. The Great Gymnasium

G. Gym Hall

H. House of the Genius

I. House of the Skeleton

J. Large Inn

K. House of the Alcoves

L. House of the Mosaic
Atrium

M. Suburban Baths

Underneath, up to 18 metres (60 feet) below the surface, they found a world where, unlike the destruction of Pompeii, everything was as well preserved as it had been over 1,800 years previously.

They found richly decorated public facilities such as swimming pools and sports grounds, wooden furniture, cloths, preserved food and charred papyrus scrolls detailing life in the ancient Roman Empire. Initially, unlike Pompeii, very few human remains were found, suggesting that most residents had made a successful escape across the sea. But eventually the discovery of the petrified skeletons of over 120 people showed this not to be true. The brain of one Herculaneum man, believed to have been a caretaker, was found to have been so incinerated by the heat that it transformed into glass, a rare instance of brain tissue being vitrified and preserved for archaeologists. Delving into the past in places such as Herculaneum can uncover far stranger offerings than we might imagine.

↑
Herculaneum was preserved beneath Vesuvius's pyroclastic flow, despite being lost for nearly 2,000 years.

Labyrinthos Caves

WHERE IS THE LOCATION OF
THE LEGENDARY GREEK LABYRINTH?

GREECE

N 35° 03′ 46″
E 24° 56′ 49″

A colossal labyrinth, deep underground. A terrifying prisoner, half man, half bull, with a name sure to chill the blood of any living creature. A cruel master, King Minos, son of Zeus, demanding fourteen youths be sent from Athens every nine years to feed to his captive. A hero, Theseus, the only one with the bravery to take on the beast and end this tyranny. A princess, Ariadne, whose ingenuity allows our hero to safely return home. And, at long last, perhaps, a real-world location to this ancient tale, where one of ancient Greece's most iconic villains met his bloody end.

Subterranea is generally opaque enough without the need to introduce further fantasy, but the Minotaur's labyrinth, wherever its mythical location truly lies — or perhaps whatever inspired the famous story — is one that really fires the imagination. The story has survived the test of time for millennia, evolving over the years, but is traceable in some form back to Homer's *Iliad* (written nearly 3,000 years ago) and quite possibly earlier.

In the early years of the twentieth century, wealthy British archaeologist Arthur Evans uncovered a ruined palace on the island of Crete — supposedly the location of the fabled labyrinth — near the town of Knossos. With a bold flair for theatrics, Evans declared this to be the palace of Minos himself, kickstarting a rumour that has now lasted over a century. Here, somewhere, he announced, were the horrifying tunnels of taurine terror where so many young Athenians met an early demise. Tourists were hooked, and over a

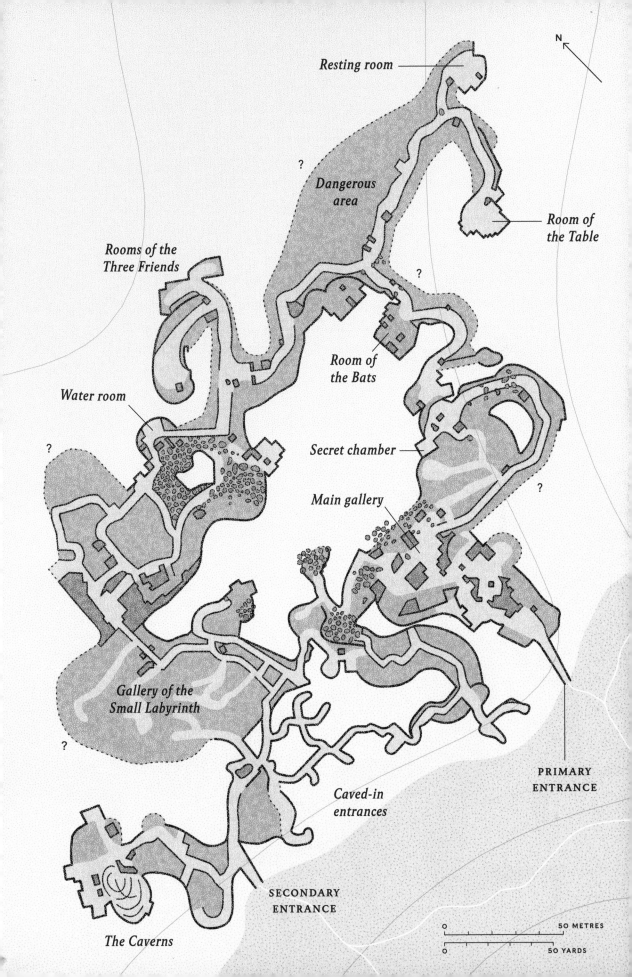

Resting room

Room of
the Table

Dangerous
area

?

?

Rooms of the
Three Friends

Water room

?

Room of
the Bats

Secret chamber

Main gallery

?

?

Gallery of the
Small Labyrinth

?

Caved-in
entrances

PRIMARY
ENTRANCE

SECONDARY
ENTRANCE

The Caverns

N

0   50 METRES

0   50 YARDS

century later the site still welcomes as many as 600,000 visitors annually, keen for a glimpse of the grand palace itself, and an opportunity to morbidly ponder the dark catacombs lost beneath their feet. Evans even made several changes to the site, including reconstructing the palace with faux-classic wooden beams and frescos to add colour and vibrancy to the attraction he had created.

But like the arrival of Theseus himself — who carried a ball of thread into the labyrinth in order to make his audacious escape after slaying the beast — this story has a twist. In 2009, an elite scholarly team composed of geographers, historians, geologists and archaeologists from the University of Oxford, working with, among others, the local Hellenic Speleological Society, announced they had followed the guidance of sixteenth-century Venetian maps, and reached the conclusion that, if there was a real-world location for the Minotaur's labyrinth, it wasn't necessarily at Knossos.

'Evans's hypothesis that the palace of Knossos is also the labyrinth should be treated more sceptically,' leader of the expedition, Nicholas Howarth, told newspapers at the time of their announcement. 'The fact that this idea prevails so strongly in the popular

imagination seems more to do with our romantic yearning to believe in the stories of the past — coupled with the power of Evans's personality and privileged position in the academy as an Oxford "don" — than with archaeological or historical fact.'

Instead, a lesser-known former quarry called the Labyrinthos Caves in Gortyn, 32 kilometres (20 miles) away, was proposed as the likelier of two possible alternatives (presumably, if nothing else, because of the name). This complex contains at least two and a half miles of tunnels, and for centuries it had been considered as a possible site for the labyrinth, before being neglected, like many others, once Evans declared Knossos to be the key location. While the experts weren't able to find hard evidence of a labyrinth, they did conclude that the tunnels had been to some extent artificially constructed and widened, a tantalising hint that they may have been built for humans to traverse.

Many of the passageways at Labyrinthos certainly look deliberately cut, almost to right angles at times. The tunnels are predominantly dry, and large enough to walk through comfortably, while many 'walls' are a mishmash of stones thrown together seemingly haphazardly. Adding to the tunnels' mythology, the cave entrances were severely damaged when ammunition being stored by Germany during the Second World War was unintentionally detonated. The remnants of these munitions continue to be a lingering threat to anyone considering wandering around the caves, even today.

Maybe this place does hold access to a labyrinth, somewhere. Maybe it's in another location entirely. Or maybe this is just a secret that the underground world will keep to itself, a legendary story that refuses to yield to contemporary empirical analysis. 'I think that each site has its claim to the mystery of the labyrinth,' continued Howarth, 'but in the end there are questions that neither archaeology nor mythology can ever completely hope to answer.'

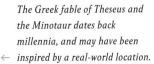

The Greek fable of Theseus and the Minotaur dates back millennia, and may have been ← inspired by a real-world location.

Tenochtitlan

MEXICO

N 19° 26' 00"
W 99° 07' 55"

Crumbling white walls, blocks of concrete crudely slammed together, rogue urban debris. This downtown street looks no different to any other in Mexico City — that is, apart from the gaping hole in the ground, as though the road surface has been hurriedly hacked away with a giant breadknife.

Rows of press reporters and photographers stand to one side, all staring down into the pit. Inside, Raul Barrera, an archaeologist from Mexico's National Institute of Anthropology and History, is holding court. And while many are indeed assiduously paying attention to his insightful words, more than the odd eye is drifting to the artefacts lined up behind him, just over his shoulder.

Understandably so, because Barrera and his team are unveiling their latest remarkable discovery — a ritual ball court and buried temple dedicated to the Aztec wind god, Ehecatl, dating all the way back to the Aztec Empire. Mere metres from the Zócalo plaza, at the very centre of the city, they were uncovered only when a neighbouring hotel decided to begin renovation work on a site that had sat undisturbed for centuries.

With 21 million people, twenty-seven skyscrapers, and twelve subway lines, Mexico City is one of the world's most epic megacities. But it's also a city with an unusual past, built directly on top of the buried remains of an ancestor. When Spaniard Hernán

Teneyocan Causeway

Tepeyacac Causeway

N

Lake Texcoco

Huixnahuac

Atepetla

Xochimanca

TLATELOCO

Great Market

AXACUALCO

Nextitlan

TENOCHTITLAN

Tlacopan Causeway

Templo Mayor

Palace of Montezuma

MOYOTLAN

ZOQUIAPAN

MIXIUHCA

Tlacateco Causeway

Zacatlamanco

Ahuehuetlan

Iztapalapa Causeway

0   3 KILOMETRES

0   2 MILES

Cortés arrived in modern-day Mexico in the early sixteenth century, he and his conquistadors were spellbound by the extraordinary Aztec capital city of Tenochtitlan they were taken to by the then-emperor Montezuma II.

Founded on two small islands in the middle of Lake Texcoco by the Mexica people in 1325, Tenochtitlan had quickly grown — as the Mexicas allied with other cultures to become the prosperous Aztec civilisation — spreading across the lake and connecting to the mainland via a series of bridges and streets. At least seventy-eight temples (possibly numbering even into the hundreds) adorned the lacustrine city streets, along with all the housing, schools and other necessities required for the estimated 400,000 resident population. 'When we saw all those cities and villages built in the water ... we were astounded,' described conquistador Bernal Díaz del Castillo in his book *The Conquest of New Spain*. 'These great towns and [prayer-houses] and buildings rising from the water, all made of stone ... some of our soldiers asked whether it was not all a dream ... It was all so wonderful that I do not know how to describe this first glimpse of things never heard or, seen, or dreamed of before.'

Official Spanish documents record that Montezuma willingly gifted Tenochtitlan and the entire empire to Cortés, and that later incidents of violence (one of which claimed the emperor's life) were simply a stout defence against guerrilla Aztec warriors. Many historians question this version of events, referring instead to grinding conflicts which eventually saw the Spanish conquer the iconic city.

Either way, once indisputably in charge of the polis, the Spanish began by founding a brand-new city in 1521, simply building their shining New World capital slap-bang on top of Tenochtitlan's existing infrastructure (in many cases using the exact same stones that the Aztecs once had). Mexico's National Palace, home to the current president, now sits atop the ancient palace of Montezuma. Furthermore, the centrepiece of the city, the grand Metropolitan Cathedral — the construction of which took almost two and a half centuries from 1573 to 1813, duly comprising a mix of

N

BASIN OF MEXICO
CIRCA 1519

Brackish water

Fresh water

Lake Zumpango

Cuautitlan

Xaltocan

Lake Xaltocan

Teotihuacan

Chiconautla

Alcoman

Sierra of Guadalupe

Ecatepec

Texcoco

Atzacoalco

Azcapotzalco

Lake Texcoco

Huexotla

Tlacopan

DIKE OF NEZAHUALCOYOTL

Chapultepec

Tenochtitlan

Mixcoac

Coyoacan

Culhuacan

Iztapalapa Peninsula

Lake Xochimilco

Lake Chalco

Xochimilco

Chalco

MEXICO CITY CIRCA 2020

—— *Mexico City urban area*

—— *Remains of Lake Texcoco*

0   20 KILOMETRES

0   10 MILES

Renaissance, Baroque and Neoclassical architectural styles — was built essentially on the same spot as the Templo Mayor, the great Aztec temple that formed the very centre of their world. In 1978, electrical workers stumbled upon an ancient monolith that, after a five-year excavation, turned out to be the decaying remnants of this once-great temple.

These days, all underground maintenance activity in the old part of the city, from water pipes to electrical cables, requires supervision by the National Institute of Anthropology and History. And as the unearthing of the Ehecatl temple demonstrates, ancient monuments continue to surface. Only a month later, another striking Aztec memorial — a tower of human skulls, men, women and children, numbering over 650 individuals — was found nearby. They are believed to be victims of sacrifices to Huitzilopochtli, god of the sun and war.

Modern Mexico City was built on top of the foundations of Tenochtitlan, centre ← of the Aztec Empire.

Some of Mexico City's treasures have recently appeared through an unfortunate phenomenon: subsistence. As might perhaps be expected when constructing a megacity in the water-logged remains of a lake basin, this historic city is sinking (heavy pumping of groundwater to supply urban demand hasn't helped). While the average rate is only around 6 centimetres (around 2.5 inches) per year, some structures are experiencing much more dramatic slides into the clay soils, up to 40 centimetres (more than a foot) annually, creating clear and obvious tilts in many heritage buildings. The aforementioned cathedral was experiencing just such a potentially catastrophic sinking prior to a large-scale stabilisation project that halted the problem.

Yet while subsistence may be a major problem for city authorities, it's a bonanza for archaeologists. As cracks appear, so opens a likely pathway to as-yet-undiscovered treasures. The use of modern technologies such as subterranean radar and 3D scanning enables the relics and ruins below these colonial monuments to become visible and analysable. In time, the history of Tenochtitlan that Cortés *et al* tried to bury for ever might yet re-emerge in full.

A tower of hundreds of skulls, believed to be victims of human sacrifices, was one monument excavated from beneath the city ← streets.

Basilica Cistern

A MAGNIFICENT SUNKEN PALACE BENEATH THE STREETS OF CONSTANTINOPLE

TURKEY

N 41° 00′ 32″
E 28° 58′ 40″

Petrus Gyllius was a scholarly man, prone to searching for answers to peculiar questions about the world around him. Upon a visit to the grand Byzantine capital of Constantinople in 1545, the curious traveller was intrigued by strange rumours that residents in this city — on the banks of the saline Sea of Marmara, between the Aegean and Black seas — were pulling buckets of fresh water from holes beneath their basements. More extraordinary, freshwater fish would sometimes appear in these buckets, with some locals even laying down a hook and line to catch dinner. It was surely an intolerable conundrum for an ichthyologist and topographer such as Gyllius. After delving into one particular basement, equipped with little more than a sketchbook and torch, he uncovered a magnificent subterranean structure hidden 10 metres (30 feet) below the city streets. The mysterious fish were seen swimming around a huge artificial lake, surrounded by hundreds of tall marble pillars stretching high to the arched ceiling above.

This impressive building was constructed in 532 AD, at the behest of Emperor Justinian I, for the purpose of storing fresh water to supply the demands of his majestic temples (such as the enormous domed cathedral of Hagia Sophia that continues to dominate the Istanbul skyline today). Thanks to the labour of an estimated 7,000 slaves, upon completion the cistern covered

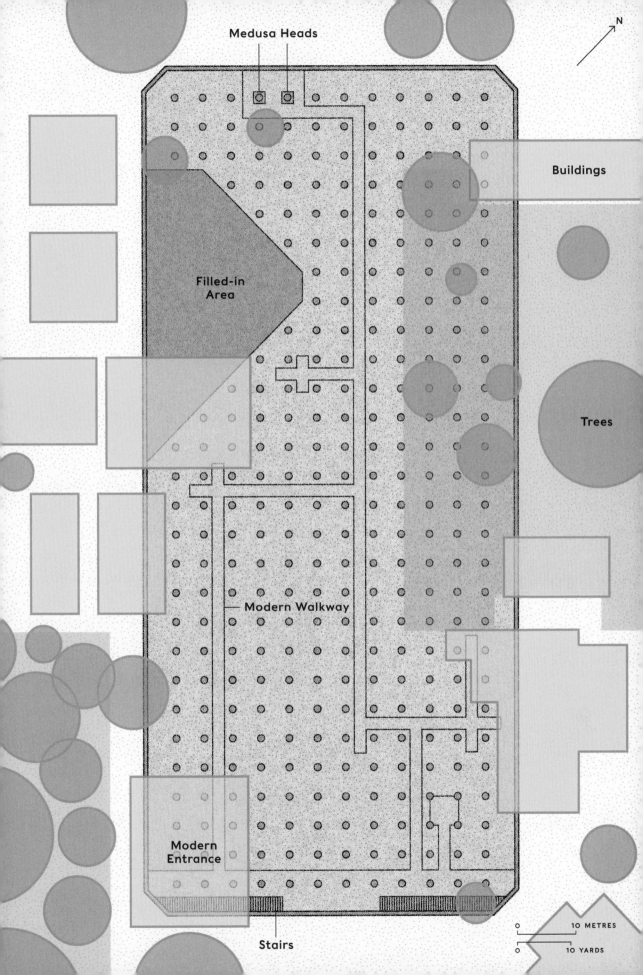

Medusa Heads

Buildings

Filled-in
Area

Trees

Modern Walkway

Modern
Entrance

Stairs

N

0   10 METRES

0   10 YARDS

Majestic marble pillars decorate the largest cisterns constructed below Constantinople, now Istanbul.

an area of 9,800 square metres (105,000 square feet) — larger than an international football pitch — sitting beneath the now-demolished Stoa Basilica. Supported by 336 columns aligned in rows, each 9 metres (27 feet) tall and built using materials recycled from the ruins of former Roman architectural marvels, the internal reservoir was filled by channelling water along aqueducts from 20 kilometres (12 miles) away. If filled to maximum capacity, it would have held something close to 80,000 cubic metres

of water, roughly thirty Olympic-sized swimming pools, almost enough to fill London's Royal Albert Hall.

The Basilica Cistern was perhaps the grandest of at least thirty cisterns built beneath Constantinople during this time, all of varying sizes, but each maintaining the same imposing style, with strong external fortifications and high domed roofs that reflect the importance the emperor placed upon this aquatic resource. With attack from invaders an ever-present threat to the city at this time, and antagonistic barbarians or opposing military forces trying to cut off the natural water supply, these cisterns enabled the flow of clean water to the residents of Constantinople to be maintained for months.

Constantinople's transition — following its eventual fall to Sultan Mehmed II and his Ottoman forces in 1453, after a nearly two-month-long siege — to becoming a modern Islamic metropolis, the centre of a thriving Ottoman Empire, saw radical change sweep the city, such as the repurposing of the city's magnificent churches into mosques. What didn't fare so well were Justinian's precious cisterns. The new inhabitants of the city were less inclined to look favourably upon such stagnant waters, and so allowed them to fall gradually into ruin. Hence, when Gyllius finally set foot inside this forgotten cistern, he found a place full of disposed debris, wild carp, and even dead bodies.

Yet the renaissance of the Basilica and other cisterns was still centuries away. Despite initial restoration attempts in both the eighteenth and nineteenth centuries, the splendid architecture was allowed to become fetid and decrepit as the surface city grew above. It took until 1985 (decades after the city had officially changed its name to the more familiar 'Istanbul') but eventually this historic subterranean layer of the city was restored, reinvigorated and, two years later, reopened to the public. It is now one of Istanbul's top tourist attractions, visitors welcomed to walk the fifty-two steps down below the surface to see 'Yerebatan Saray' — the 'sunken palace', as it is known in Turkish.

Among the most revered treasures contained within are the stone heads of Medusa — the snake-haired Greek monster capable of turning people to, ironically, stone — at the base of two columns. One of the Medusas lies on her side, while the other is completely upside down, a reminder that the cistern builders saw these Roman raw materials as little more than meaningless, interchangeable building blocks. The cistern has even taken to the silver screen, as the backdrop to high-budget Hollywood blockbuster adaptations such as James Bond's *From Russia with Love* and the Dan Brown thriller *Inferno*. Quite the turnaround for a hidden reservoir, lost to the world for centuries.

The upturned head of Medusa serves as the base of two high-profile pillars.

→

Elephanta Caves

AN UNDERGROUND TEMPLE FILLED
WITH EXTRAORDINARY SCULPTURES

INDIA

N 18° 57' 45"
E 72° 56' 00"

It's only one line, but then anything written by the original Renaissance man, Leonardo da Vinci, is worthy of being pored over by experts for centuries. 'Map of Elephanta in India which belongs to Antonello the merchant' he once scribbled in one of his famous notebooks. The identity of the aforementioned Antonello remains an unsolved mystery, but, whoever he was, it seems he captured the remarkable Italian thinker's interest with his cartographic information from India. This was a country whose accessibility from Europe during this time — the early days of the sixteenth century — was greatly enhanced by the recent discovery of a sea route, paving the way for ambitious tradespeople from Florence and Milan to begin travelling back and forth in search of riches.

The 'Elephanta' in question, however, we are well informed about. It refers to a remarkable collection of handcrafted cave temples on Gharapuri, one of a handful of harbour islands around 10 kilometres (6 miles) off the coast of the Indian city of Bombay, since renamed Mumbai. First inhabited around the second century, the temples were carved from the basalt hillside somewhere between the fifth and eighth centuries by Hindu monks, paying tribute to Shiva, a central deity in Hinduism who acts as god of both creation and destruction. Within the 5,000 square metre (54,800 square foot) complex, the main sacred temple consists of a long hall laid out in the shape of a mandala, featuring various detailed sculptured panels on the walls. A three-faced Sadashiva

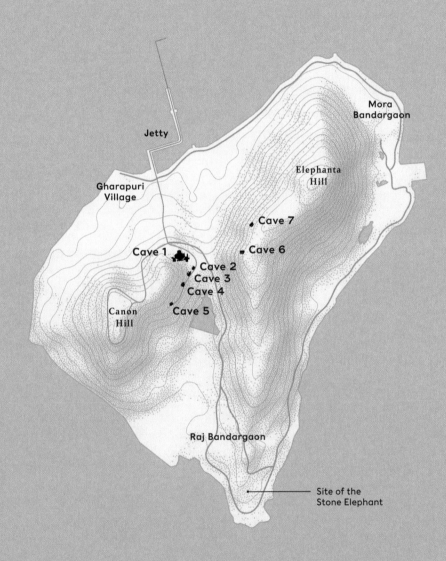

N

Mora
Bandargaon

Jetty

Elephanta
Hill

Gharapuri
Village

Cave 7

Cave 1

Cave 6

Cave 2

Cave 3

Cave 4

Canon
Hill

Cave 5

Raj Bandargaon

Site of the
Stone Elephant

A r a b i a n   S e a

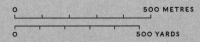

0   500 METRES

0   500 YARDS

statue, positioned inside the cave temple, exhibits both of Shiva's dual, polarising identities (plus a third, neutral face). It's a manifestation of the central lesson within Hinduism, that life is a cycle. Destruction follows creation, and vice versa.

When Portuguese colonisers arrived in the sixteenth century, they gave the island the name 'Ilha Elefante' (Elephant Island) in reference to an impressive, life-size stone elephant sculpture erected there. The elephant in question was later relocated to Bombay, although thanks to the clumsiness of the removal crew, it required significant surgical repairs after being broken apart en route. It now sits, visibly re-welded and somewhat degraded, in Jijamata Udyan, a public garden in the renamed city of Mumbai.

The unfortunate fate of the elephant is but one example of a succession of incidents of cultural vandalism in the temples by Portuguese settlers during the occupational years, historians

The remarkable temples of Elephanta were sculpted into the basalt bedrock of the island.
↓

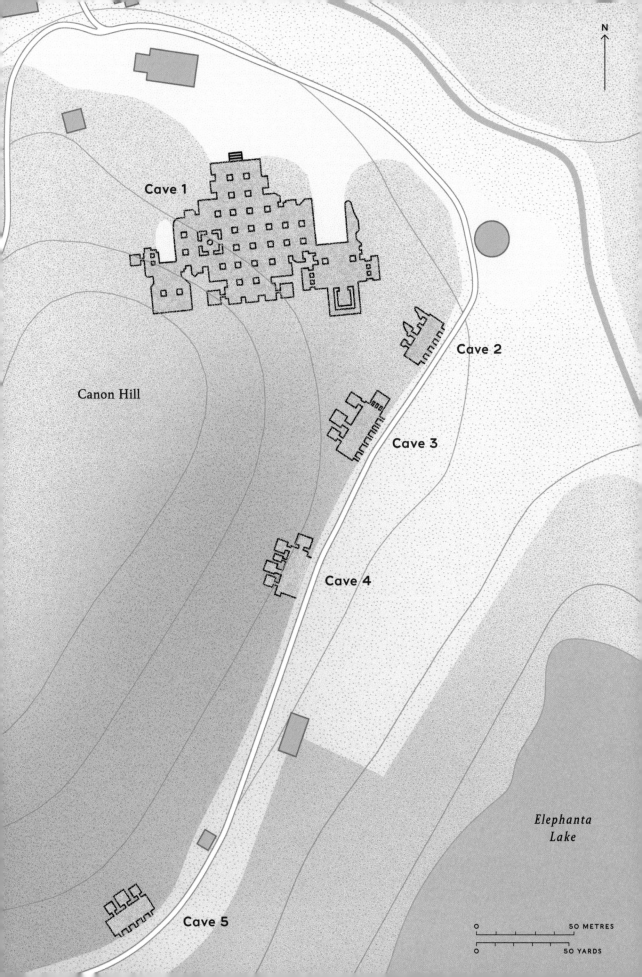

N

Cave 1

Canon Hill

Cave 2

Cave 3

Cave 4

Elephanta
Lake

Cave 5

O   50 METRES

O   50 YARDS

claim. It is believed that, upon taking control of the island in 1534, they broke off parts of many statues for little more than their own amusement, defaced once-revered monuments, and stole a symbolically important inscribed stone that has since been completely lost to history. While the remaining statues remain a popular and impressive sight, in reality they are widely believed to be a shadow of what would have been found here half a millennium ago.

By the late twentieth century, the decision was finally made to clean up the caves, in an attempt to restore them and the temples within to their former grandeur. While for many international visitors, Elephanta Island and its impressive caves are a picturesque tourist attraction, for many of India's more than half a billion Hindus, it is perhaps more akin to a small-scale pilgrimage, a trip to a place not only of worship, but also education. To teach about faith, as the original constructors likely intended.

Unfortunately, the heavy rains and subsequent flooding that have deluged the island in recent years are perhaps a sign of things to come. As if hundreds of years of chronic neglect wasn't enough, UNESCO has identified Elephanta as one of over a hundred World Heritage sites most threatened by the impact of rising sea levels in the coming years, due to its precarious position on the edge of the Arabian Sea. After creation comes destruction, and the cycle continues.

CAVE 1
The Main Cave

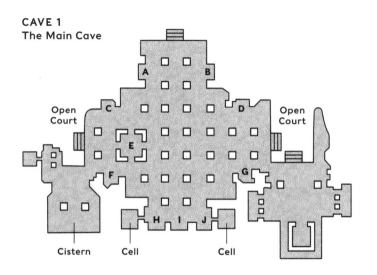

Monuments

A. Nataraja Shiva

B. Shiva as Lakuliśa

C. Andhakasurvadhamurti Shiva

D. Ravana under Kailasa

E. Shiva Shrine

F. Kalyanasundaramurti Shiva

G. Parvati in Attitude of Mana

H. Gangadhara Shiva

I. Mahadeva

J. Ardhanarisvara Shiva

Qumran Caves

HIDING PLACE OF THE DEAD SEA SCROLLS

ISRAEL

N 31° 44′ 43″
E 35° 27′ 33″

On the northwest shore of the Dead Sea, an excavation brings a potential new theological treasure up to the light. Here, in this region of Qumran in 2017, 'Operation Scroll' is underway, an effort to locate caves unknown to academia, and potentially find many valuable objects within. The excavation has already discovered one cave home to many flint blades and arrowheads dating back to Neolithic times, suggesting it was first inhabited at least 4,000 years ago, and potentially over 10,000. But these aren't the excavators' primary target. Instead, attention has been grabbed by the discovery of a single jar, with an intact scroll of parchment inside. Amid excitement, this specimen is rushed to the nearby Hebrew University for analysis. Could this be the biggest archaeological find in this region for over sixty years?

At first glance, there is little to indicate that the Qumran caves are unique or particularly significant to history. Small cracks in rocks, buried out of sight beneath a landscape littered with many other small cracks in rocks, they consist primarily of dark, dry, barren spaces existing in a world where time moves almost unfathomably slowly, the pace of life drifting past like desert dust caught in a warm breeze. Only the most imaginative passer-by would be likely to guess at the international attention these mundane caves would attract in the middle of the twentieth century.

N

Cave 3

Cave 11

Cave 1

Cave 2

Kalya
Kibbutz

Qumran

Cave 6

Cave 5

Caves 7–9

Cave 10

Cave 4

Nature Reserve

*D e a d
S e a*

Cave 12

0   500 METRES

0   500 YARDS

In 1947, Bedouin shepherds were passing through Qumran, searching for a lost sheep. Attracted to a cave entrance, they stepped inside, to be greeted by a large number of strange objects, including papyrus parchments bound by leather and copper. Upon the announcement of their discovery, a major excavation operation began in 1949, which, over the following decade, identified eleven different caves in the Qumran region believed to have once held similar parchments. However, the excavation arrived too late. By the time experts were on the scene, most of what they found was evidence of how thoroughly the caves had been ransacked. While some remaining objects — such as pottery and empty jars — were salvaged and taken away for study, it was plain that the sites had been thoroughly looted, and many invaluable items taken.

While to archaeologists, the lost artefacts were priceless, this

↑

The eleven famous caves are sprinkled across the rugged and mountainous region of Qumran.

clearly wasn't the case for whoever took them and put them up for sale on the black market. Once it became apparent that the texts were available to purchase, scholars began arranging for the return of the loot. In this way, upwards of 15,000 fragments of texts, from nearly 1,000 manuscripts, have been recovered over the intervening years. Now primarily on display in the Israel Museum, they are commonly known as the Dead Sea Scrolls, and are considered some of the most important and best-preserved texts ever recovered from the ancient world.

Together, the scrolls include some of the earliest known biblical texts, symbolically and historically important to both Christianity and Judaism. They are believed to have been hidden in the caves by the Essenes, a splinter sect of Palestinian Jews who prospered in Qumran around 2,000 years ago, the parchments also revealing much about their own unique lifestyles and worldviews. While Israel has claimed ownership of the scrolls to date, there are ongoing rival challenges on behalf of both the Palestinians (who contest much of what Israel believes to be its territory) and Jordanians (on the other side of the Dead Sea), who argue that their long history in the region entitles them to share in the cultural heritage the Dead Sea Scrolls represent.

Analysis of the parchments themselves has revealed that they contain quantities of sulphur, sodium and calcium, salts believed to have helped preserve the documents for nearly 2,000 years, far longer than might otherwise have been expected. The habit of the Essenes in hiding their manuscripts underneath bat guano and other cave debris (to discourage thieves, unsuccessfully as it turned out) may also have helped extend their lifespan. In some cases, fragments believed to be from the same parchment have later been revealed to be made of the hides of completely different animals, such as sheep or cow, further complicating the colossal jigsaw puzzle of how all the fragments fit together. These findings have also helped identify genuine scrolls, and prevent deliberate forgeries from slipping unnoticed into the collections.

All the more reason for the excitement surrounding the latest parchment. After years of missed opportunities, and decades of

hoping to find a fresh haul of documents, could this be it? Could this newly discovered scroll shed more light on the ancient world? Is this perhaps an historic twelfth Qumran cave?

The results were disappointing. The parchment was genuine, dating back around 2,000 years, just like the original scrolls. But it was completely blank. Experts believe it was left in the cave in preparation for writing, but the words were never added. Although, if they had been, perhaps it would never have been found — the presence of many broken jars, leather straps and wrapping cloths (as well as a pair of modern iron pick-axe heads) suggests that looters also discovered this cave in the 1940s or 50s, and took almost everything of value. The blank parchment in the jar was one of the few objects left behind. While disappointing, it nevertheless confirms that scrolls were left in more than just the original eleven caves. The Qumran caves may have many more stories yet to tell.

↑
The Dead Sea Scrolls, fragments of parchment recovered from the Qumran caves, are considered priceless artefacts from the ancient world.

Cueva de los Tayos

A MYSTERIOUS CAVE, THEORISED
TO HAVE BEEN BUILT BY AN ANCIENT
CIVILISATION, AND CONTAINING THE
ORIGINS OF HUMANITY

ECUADOR

S 04° 18′ 27″
W 78° 40′ 53″

A grand cave passageway, dug into the side of the Andes, deep in the Amazon rainforest. A strangely square opening in the rock, surely far too quadratic to be have formed naturally. Such an entrance must have been created intentionally, probably by a long-lost civilisation. At least, that was the theory this 1976 expedition had come here, to remote Ecuador, to investigate. With the blessing of their honorary president, the famous astronaut Neil Armstrong — who just a few years earlier had walked on the surface of the moon — they tentatively entered the cave in question, eager to solve a mystery that had captured the imagination of the world.

Their reason for being here, thousands of miles from home, can be traced back to one man: the writer Erich von Däniken. After success with his earlier published work *Chariots of the Gods?* — a book that leant heavily on the existence of extraterrestrial life coming to Earth from outer space — the Swiss writer was continuing his literary delve into the supernatural. His next book, *The Gold of the Gods*, was dedicated to the story of Hungarian-Argentine speleologist János Juan Móricz's 1965 journey to southeast Ecuador, near the border with Peru, to what was known as the Cueva de los Tayos ('Cave of the Oilbirds'). He wrote that Móricz had discovered remarkable sculptures inside the cave, including a sensational library consisting of large metal sheets engraved with signs and words that told the origin story of

↤ The horizontal layers of rock in the Móricz arch have convinced many visitors that this cave is undoubtedly a man-made creation.

humanity. Däniken stated he too had later seen these objects with his own eyes.

Published in 1972, the book's reception was immense. The desire to learn more about this strange cave of treasures buried in the wilderness was intoxicating, something of a modern-day quest for El Dorado. There was a scramble among explorers and adventurers to assemble the necessary people, equipment and, of course, funding, to get a full scientific expedition off the ground. Eventually, it was Stan Hall, a civil engineer in Scotland, who dedicated himself to making the idea into a reality. He assembled a British-Ecuadorian group of over a hundred — including scientists affiliated with reputable British institutions such as the University of Edinburgh and the British Museum, plus government forces and military personnel — to find out more about the undiscovered ancient civilisation that was apparently responsible for the remarkable objects inside the cave. Given it was only seven years since the iconic, era-defining NASA *Apollo 11* moon landing, it was quite a coup for Hall to get Neil Armstrong, first human to set foot on the lunar surface, to be the expedition's honorary president, and lead them into the unknown.

In July 1976, after several days travelling from the capital Quito, the group arrived at the cave in question. After initially

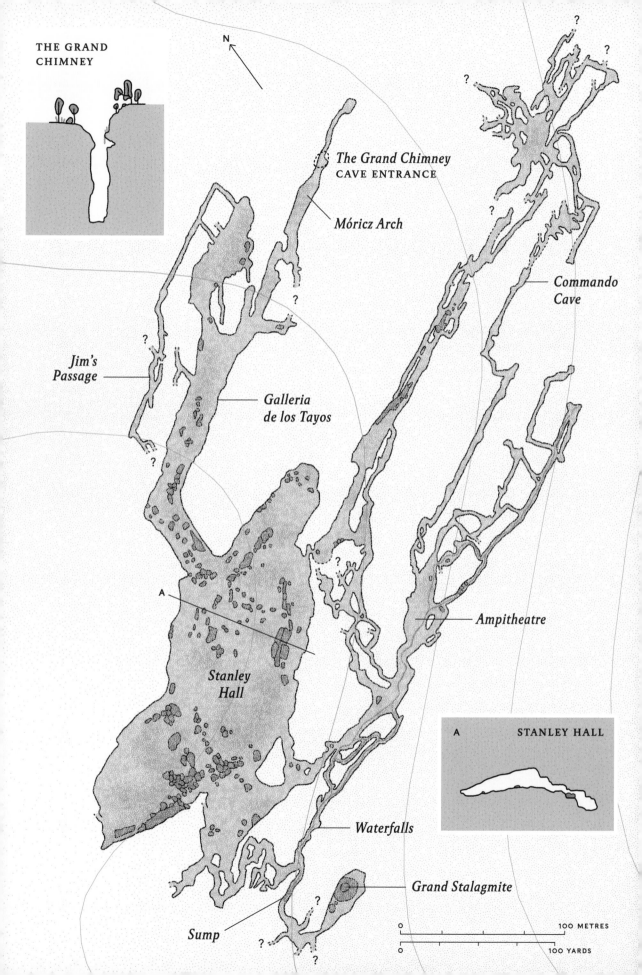

THE GRAND
CHIMNEY

N

The Grand Chimney
CAVE ENTRANCE

Móricz Arch

Commando
Cave

Jim's
Passage

Galleria
de los Tayos

Ampitheatre

Stanley
Hall

A   STANLEY HALL

Waterfalls

Grand Stalagmite

Sump

0   100 METRES

0   100 YARDS

plunging into the earth down a vertical shaft over 60 metres (200 feet) deep, they passed through a large passageway with a cathedral-like ceiling high above their heads. Navigating a series of narrow passageways decorated with stalactites and stalagmites, they located the famous square entrance — now often referred to as the Móricz Arch. Pushing on from this point, they went on to encounter vast flooded caverns, and wildlife including rainbow boas, tarantulas and the nocturnal *tayos* birds for whom the cave is named after.

What they didn't find, however, were any metal sheets, any documented history of humanity, or indeed any evidence of treasures from a past civilisation. Däniken was unimpressed, and wrote to Armstrong to complain that the expedition must be mistaken — perhaps they explored a different location by accident? — but the famous astronaut was unmoved, and the expedition formally concluded.

While the internet (and other less scrutinised spaces of debate) might still play host to many claims of irrefutable proof that Cueva de los Tayos was made by human hands, to date the scientific world has failed to find any evidence to support them. The truth may be somewhat deflating. It seems that while the cave may at some stage in history have been inhabited by members of the indigenous Shuar people, for whom it is certainly an importantly spiritual place utilised for traditional ceremonies, the shapes of the formations themselves are merely a product of the underlying geology. Sandstone is certainly capable of creating strangely flat and smooth rocks that give the appearance of being sculpted by hand.

A fresh expedition, orchestrated by Stan Hall's daughter Eileen and the UK-based Open Close art collective in 2018, has richly documented the cave in the form of photography, sound recordings, video and 3D mapping, so as to better investigate the strange interior spaces. While some may be interested in looking into the claims of Móricz and Däniken, for others, promoting the cause for protecting the cave — believed to be the largest in Ecuador — and possibly even making it a UNESCO World Heritage site, is perhaps sufficient motivation for further exploration.

A dramatic waterfall of light illuminates the base of the sheer shaft that visitors to Tayos must descend in order to navigate the deep cave interior.

→

Modern History

London Underground

**UNITED
KINGDOM**

N 51° 31' 14"
E 00° 05' 58"

Tools in hand, archaeologists delicately brush aside earth, taking extra care not to damage the fragile items they're exposing. Working in a shaft 5.5 metres (18 feet) in diameter — part of a new underground construction site in the Farringdon area of east London — they have just uncovered a familiar object only a few metres below the road. They carefully clear away more earth. Now it is unmistakable. A skeleton. They keep digging. Another skeleton. And another. Then many more.

Many major cities have their murky and mysterious pasts buried beneath layers of shiny new streets and modern buildings, but few more so than London. Here, the aura of 8,000 years of history still wafts through the air. It's certainly carried within the headlines of unexploded wartime bombs and enormous 'fatbergs' found trying to squeeze waste from twenty-first-century London through the city's ancient sewerage system.

The global practice of burrowing underneath the city for the purpose of public transportation was pioneered here; the construction of an underground railway line between Bishop's Road (now called Paddington) and Farringdon Street in January 1863 was the first step in the making of the world-famous 'London Underground'. As well as giving us the iconic roundel red-circle-with-blue-strikethrough logo, Harry Beck's classic schematic tube map, and the droning order to 'mind the gap', the Underground would go on to spawn eleven lines and 270 active stations over

400 kilometres (250 miles), now capable of carrying 5 million daily passengers, or as many as 1.35 billion annually.

A persistent rumour floats around Londoners that many of the strangely curved Underground lines of yesteryear were constructed specifically to avoid the locations of mass graves. Probably untrue, but nevertheless perhaps something to be considered by the planners of Crossrail, the newest large-scale addition to London's mass-transit system. Historians had for centuries noted their belief that Farringdon contained a so-called 'no man's land' mass grave. It's said to be where many victims of the infamous Black Death — the plague that arrived on Britain's shores in 1348 and wiped out between one third and a half of the national population — had been unceremoniously buried in their thousands. Somehow, it had always evaded detection.

But when the full March 2013 discovery of twenty-five skeletons, lined up in two neat rows, had been excavated from the Farringdon Crossrail site, the preliminary evidence suggested that this morbid resting place was finally being uncovered. When the forensic results came in, the hypothesis was proved correct. These individuals had died because of *Yersinia pestis*, the bacteria associated with the Black Death. Thankfully for everyone involved in the excavation, over six centuries is more than enough time for the bacteria itself to perish in the soil.

Any major construction beneath London runs the risk of unearthing hidden medieval burials.
→

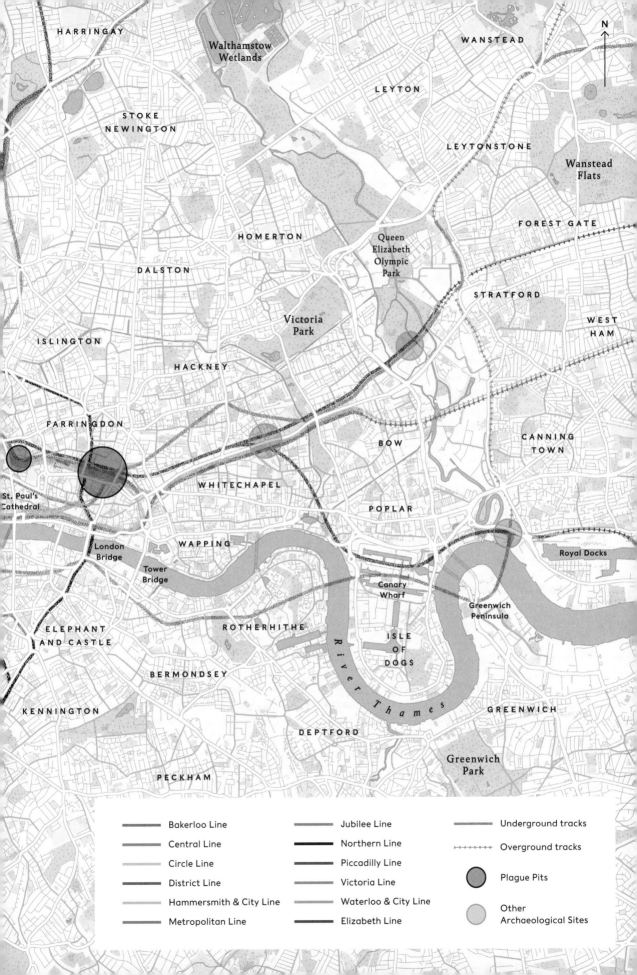

N

HARRINGAY

Walthamstow
Wetlands

WANSTEAD

STOKE
NEWINGTON

LEYTON

LEYTONSTONE

Wanstead
Flats

FOREST GATE

HOMERTON

DALSTON

Queen
Elizabeth
Olympic
Park

STRATFORD

ISLINGTON

Victoria
Park

WEST
HAM

HACKNEY

FARRINGDON

St. Paul's
Cathedral

BOW

CANNING
TOWN

WHITECHAPEL

POPLAR

London
Bridge

WAPPING

Royal Docks

Tower
Bridge

Canary
Wharf

Greenwich
Peninsula

ELEPHANT
AND CASTLE

ROTHERHITHE

ISLE
OF
DOGS

BERMONDSEY

River Thames

GREENWICH

KENNINGTON

DEPTFORD

Greenwich
Park

PECKHAM

	Bakerloo Line		Jubilee Line		Underground tracks
	Central Line		Northern Line		Overground tracks
	Circle Line		Piccadilly Line		Plague Pits
	District Line		Victoria Line		
	Hammersmith & City Line		Waterloo & City Line		Other Archaeological Sites
	Metropolitan Line		Elizabeth Line		

But this was no isolated incident. Two years later, more skeletons were found less than a mile away — or just one stop along the Hammersmith and City line — in more tunnels being dug out beneath Liverpool Street station. This time, they turned up in their thousands. Skeletons of the old, the young, the rich, the poor, men, women, from all over the capital, all apparently bundled in together.

Upwards of 3,000 skeletons were eventually extracted from this site, believed to be the old Bedlam cemetery that was used as a burial ground for inmates of the nearby Bethlehem Hospital for the mentally ill (the word 'bedlam' derives from this institution). Written records claim that this became the final resting place for at least 20,000 individuals during this era. All these people were linked by their place on the outskirts of society, or simply their inability to pay for a proper Christian burial. Some were victims of violence, some were executed. But many perished from that other horrifying wave of illness and death that rampaged through the capital's streets — the Great Plague of 1665, which killed over 100,000 Londoners. Once again, *Yersinia pestis* was the culprit. These seventeenth-century skeletons, dateable in part to the artefacts such as pottery, glass and coffin handles found alongside them, caught the bacteria red-handed.

Eventually, over 10,000 artefacts were unearthed during Crossrail's seven-year construction, from forty different locations along the line. As the project unfolded, 200 archaeologists were employed for seven years on what (perhaps inadvertently) became the largest archaeology project ever undertaken in the UK. The objects they extracted from the soil tug hard on the outermost threads of the timeline of British history, encompassing everything from prehistoric flints to Roman horseshoes, medieval animal bones to a Tudor-era bowling ball. A truly anthropomorphic layer of geology, compressed hard into the ground, before being retrieved and analysed in the twenty-first century.

Even as Crossrail's shiny new platforms begin operation, heralding the dawn of a new modern era for the big smoke, the powers that be shouldn't be complacent. There are many stations across the city that have failed the test of time. Forty-nine such abandoned stations (not all underground) can reportedly be found on the network, the clatter of shoes and the rush of passing trains merely a distant memory. Some, such as King William Street and York Road, sit gathering dust. Others have since taken on a new lease of life.

Down Street is one such example, a temporary Second World War bunker prior to the construction of Churchill's famous War Rooms (and now a 'secret London' tour destination). The old Aldwych station is now a busy filming location. Brompton Road station, near glitzy Harrods, was once an important Ministry of Defence facility, before being sold for £53 million in 2014, to developers supposedly keen to turn it into a luxury property complex. There's even a working hydroponics farm in a disused tunnel near Clapham North, with rocket, broccoli, garlic, chives, mustard leaf and edible flowers growing inside two and a half hectares of what was also once an air-raid shelter.

In London, it can be very hard for the past to remain in the past, as thousands of disturbed skeletons now know only too well.

Even today, there are many abandoned stations on the network. Some – such as Down Street, once an active station on the Piccadilly Line – have taken on second lives, while others lie dormant.
↓

Tunnel 57

LOCATION OF A MASS ESCAPE, THROUGH A TUNNEL UNDER THE BERLIN WALL

GERMANY

N 52° 32' 12"
E 13° 23' 36"

Saturday night in East Berlin, October 1964. A small group of young people walk nervously but purposefully through the streets. They try to appear inconspicuous, ignoring their thumping heart-beats, so as not to attract the suspicion of the border guards who patrol the area. Their eyes dart around as they count down the house numbers. Reaching their destination, building number 55, they quickly step inside, away from prying eyes. 'Tokyo,' they whisper, to the people waiting inside. A nod, and, after removing their shoes, they are led through a hallway and into the courtyard outside. Moments later, the first of the group is on his belly, staring down a tiny, intimidating tunnel. A tunnel that will lead them all to freedom.

Joachim Neumann had been a student in the East in the early 1960s. He had personally escaped to the free West, part of the Federal Republic of Germany, by pretending to be a Swiss citizen. But the ongoing dangers faced by the family and friends he'd left behind drove him and his peers to consider an audacious plan: tunnelling beneath the infamous Berlin Wall — constructed three years earlier, nearly 4 metres (12 feet) high, 155 kilometres (100 miles) long, driving a barricade through the heart of the capital — to enable many more people to make their escape.

In spring 1964, the plan was put into motion. Starting in the basement of a disused bakery located near the wall, Neumann and the rest of the conspirators scratched away at the earth, their

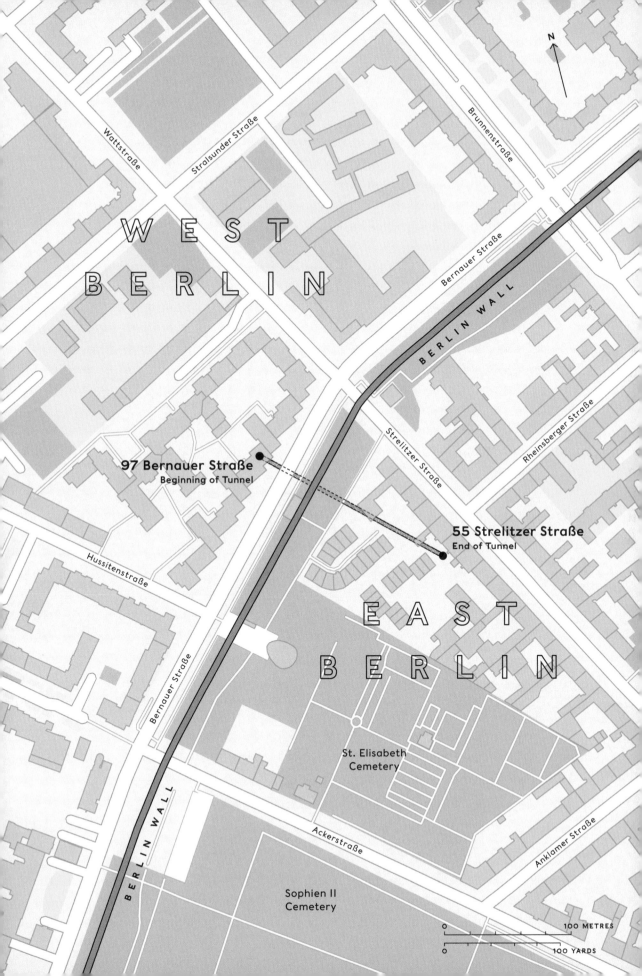

N

WEST
BERLIN

Wattstraße

Stralsunder Straße

Brunnenstraße

Bernauer Straße

BERLIN WALL

Strelitzer Straße

Rheinsberger Straße

97 Bernauer Straße
Beginning of Tunnel

55 Strelitzer Straße
End of Tunnel

Hussitenstraße

EAST
BERLIN

Bernauer Straße

St. Elisabeth
Cemetery

BERLIN WALL

Ackerstraße

Anklamer Straße

Sophien II
Cemetery

0   100 METRES
0   100 YARDS

small garden spades and a rudimentary wagon for lugging out soil their only proper equipment.

Dirt, darkness. The stench of sweat and fear. Fear of collapse, of imminent death or slow suffocation underneath enormous piles of earth. But also fear of making too much noise, of being heard by listening guards, of potential capture and execution. After twelve hours, one shift would end, and another would begin. Day after day, night after night, taking turns to make as much progress as possible. All living on site for weeks on end, so as to avoid suspicion about a steady stream of people arriving and departing

↑
The Berlin Wall separated East and West for twenty-eight years. Tunnel 57 saw the largest single escape of East Berliners beneath it.

each day. After going straight down until they hit the water table, they switched to tunnelling horizontally. This carried them slowly underneath what was known as the terrifying 'Death Strip', equipped with steel spikes and floodlights.

Eventually, after five months, a breakthrough was made. Against the odds, given the blindness that comes with secretive subterranean burrowing, they had the good fortune to break through into a disused outhouse behind an apartment building. The tunnellers had just dug their way into East Berlin, a city brutally controlled by the authoritarian German Democratic Republic.

Word soon spread. For two nights, 55 Strelitzer Straße welcomed nonchalant-looking visitors, who all made their escape through this underground passageway to a new life in West Berlin. With the tunnel no more than 80 centimetres (32 inches) wide — and even lower in height — escapees needed to scramble along on their stomachs, slowly pulling themselves through. After 140 metres (460 feet), they eventually reached the safety of the abandoned bakery. They had crossed to the other side of the wall, and were finally free.

However, late on the second night, soon after Neumann had managed to get his girlfriend (who had, coincidentally, just been released from jail) through the tunnel, some guards finally decided to investigate where all these people were going. They quickly discovered what was going on and sounded the alarm. The plotters fled, shots were fired, one guard was killed, and the entire project was halted. A total of fifty-seven East Berliners made their escape over the two nights, hence the name 'Tunnel 57'.

The Berlin Wall would continue to stand until its famous destruction in November 1989. Many escape attempts were made during the nearly three decades in which it stood — Tunnel 57 was neither the first not the last. While only a handful were successful, such as 'Tunnel 29' (an NBC TV crew even filmed the escapes as they were happening), none freed more people through a single effort than the endeavour led by Neumann — a significant proportion of the nearly 300 people who escaped underneath the Berlin Wall.

Burlington

A SECRET BUNKER TO HOUSE THE
BRITISH GOVERNMENT IN THE EVENT
OF A NUCLEAR STRIKE

**UNITED
KINGDOM**

N 51° 25′ 11″
W 02° 13′ 14″

*Burlington was built
to ensure a post-nuclear
UK government could
still perform basic
← responsibilities.*

On 13 September 1961, in the early days of the Cold War, British Prime Minister Harold Macmillan received a memo with an apocalyptic tone. In the event of a nuclear attack, members of the government would be taken to different parts of the country. In this way, the survival of at least some ministers capable of governing the nation (or returning fire, if mutually guaranteed destruction be the tactic of choice) should be assured. Therefore, if such an emergency were to befall the country, Macmillan himself would stay in London, accompanied by his Foreign Secretary, Minister of Defence and many other high-profile figures ready and willing to push the red button if it were deemed necessary. However, an equally senior cast of government ministers — including the Chancellor of the Exchequer, First Lord of the Admiralty and the Secretary of State for War — were to be quickly removed from the city and taken to an unlikely location in a remote corner of north Wiltshire, in the southwest of England.

Golf courses. Fields of livestock. Gently winding roads. On the surface, this area of Wiltshire looks much like any other rural part of Britain, green and mundane. But although there is no mention of it on road signs, underground lies a top-secret facility kept classified for decades. 'This is a prohibited place within the meaning of the Official Secrets Act' reads a warning sign. 'Unauthorised persons may be arrested and prosecuted.' Below

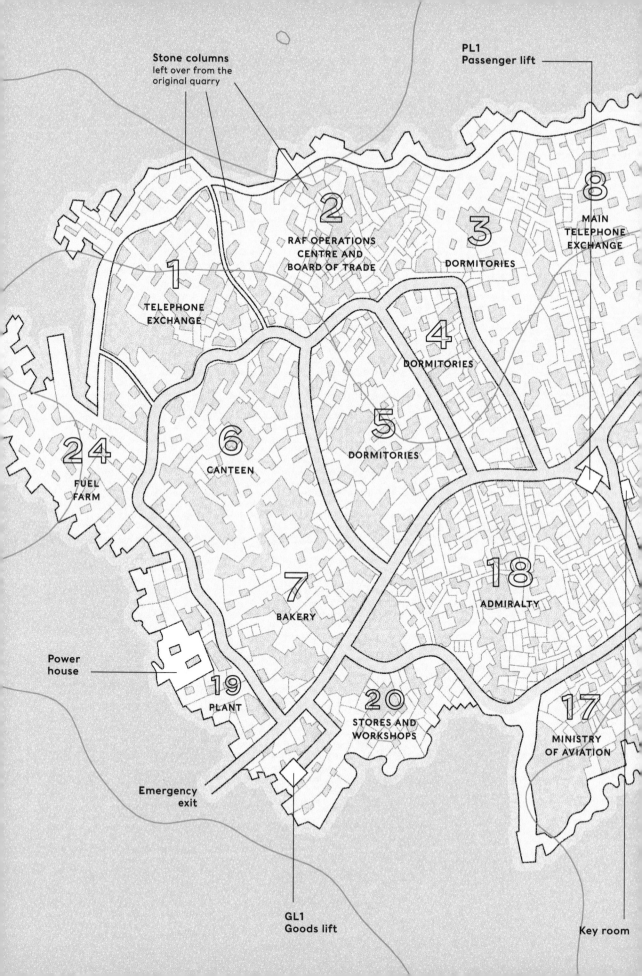

Stone columns
left over from the
original quarry

PL1
Passenger lift

2
RAF OPERATIONS
CENTRE AND
BOARD OF TRADE

3
DORMITORIES

8
MAIN
TELEPHONE
EXCHANGE

1
TELEPHONE
EXCHANGE

4
DORMITORIES

5
DORMITORIES

24
FUEL
FARM

6
CANTEEN

18
ADMIRALTY

7
BAKERY

Power
house

19
PLANT

20
STORES AND
WORKSHOPS

17
MINISTRY
OF AVIATION

Emergency
exit

GL1
Goods lift

Key room

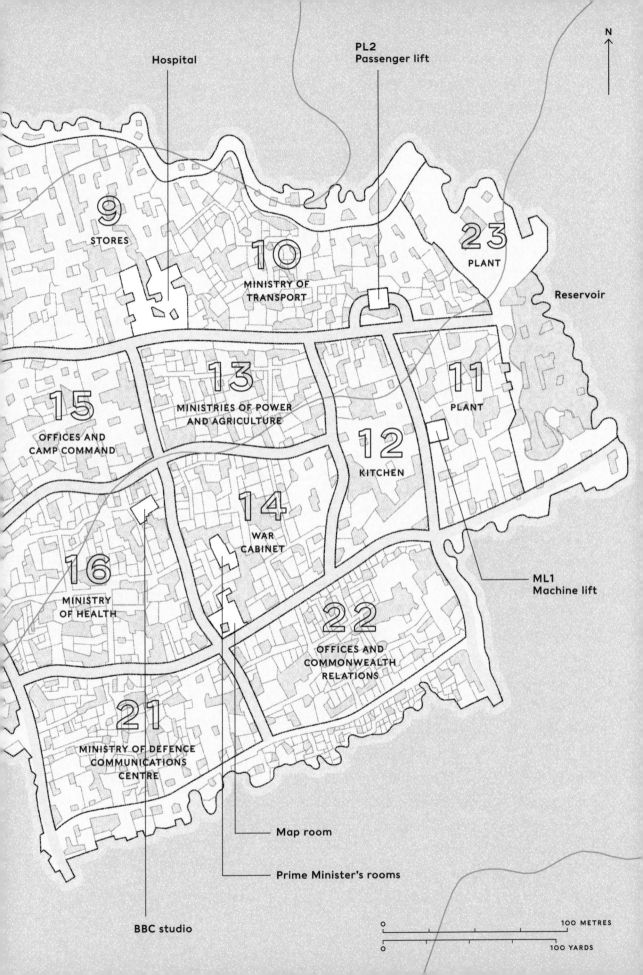

N

Hospital

PL2
Passenger lift

9
STORES

10
MINISTRY OF
TRANSPORT

23
PLANT

Reservoir

15
OFFICES AND
CAMP COMMAND

13
MINISTRIES OF POWER
AND AGRICULTURE

11
PLANT

12
KITCHEN

14
WAR
CABINET

ML1
Machine lift

16
MINISTRY
OF HEALTH

22
OFFICES AND
COMMONWEALTH
RELATIONS

21
MINISTRY OF DEFENCE
COMMUNICATIONS
CENTRE

Map room

Prime Minister's rooms

BBC studio

0   100 METRES

0   100 YARDS

ground, a strange and subversive installation: a subterranean bunker fortified with thick reinforced concrete walls, blast doors and ventilation shafts.

At times this place was called 'Stockwell', or 'Turnstile', but the codename that has gone down in history is 'Burlington'. Constructed in the latter days of the 1950s, in the event of the Cold War suddenly heating up, over 4,000 people — primarily government ministers and civil servants — were capable of taking shelter here, 36 metres (120 feet) underground in the converted remains of an old Bath stone quarry near the town of Corsham. The mine had ceased operations in 1940, and briefly been prepared as the construction site for an underground aircraft factory in the latter half of the Second World War, before falling quiet.

The design of this potential War Headquarters closely mirrored that to be found in the halls of government in Westminster, albeit with more breeze blocks and scattered rocky columns left over from mining days. It was a purely functional site, containing only the facilities deemed most essential, including a map and control room, a canteen, hospital, dentist, launderette, an underground reservoir for drinking water and four large generators supplying the site's very own power station. The second largest telephone exchange in the country was installed here, allowing ministers to maintain communication with the outside world, while a fully equipped BBC television studio would have enabled senior government officials to broadcast messages of hope and support to a battered nation, presumably cowering beneath makeshift forts and battling the effects of lingering radiation.

Close followers of contemporary history will have noticed that nuclear war failed to materialise, and in 1989 the cost of maintaining the site (supposedly as much as £40 million) led to the decision to shut down the facility. Historic England, a public body assigned with protecting historic English monuments, describes Burlington as 'an unparalleled example of our national Cold War defence heritage'.

Many of the names on the memo that passed Macmillan's desk would have left their posts without ever knowing that this

↑
Sixty years since construction, much of the site remains frozen in time.

place existed, let alone that there was a plan to bring them to live here long term. When the site was finally declassified in the early twenty-first century, visitors reported seeing many of the three months' supply of essentials like food, water and toilet paper waiting to be unpacked. Chairs, pinboards and 1960s technology such as teleprinters were still on display. Unopened, stained, heavily decaying — all waiting for a nuclear winter that thankfully never came.

Coober Pedy

**AN ENTIRE SUBTERRANEAN TOWN,
BUILT TO HELP RESIDENTS BETTER
COPE WITH AN EXTREME CLIMATE**

AUSTRALIA

**S 29° 00′ 50″
E 134° 35′ 16″**

Over a hundred years ago, on a scorching-hot summer day, a fourteen-year-old boy was walking through the red dust of a seemingly desolate Australian desert when he spotted a sparkling rock among the detritus of the terrain. Willie Hutchison and his father had travelled here by camel on the hunt for gold — which this rock certainly wasn't. But its iridescence caught Willie's attention, nevertheless. It was an astute observation. The young Hutchison had just stumbled upon opal, another valuable mineral that would utterly transform this remote spot of central South Australia.

As word of his discovery spread, aspiring miners rushed to the remote spot, an 850-kilometre (528-mile) journey inland from the coastal state capital, Adelaide. Within five years, the influx had resulted in the founding of a functioning settlement named Coober Pedy (a corruption of the local Aboriginal phrase *kupa piti* — 'white man's hole'). The economic downturn of the 1930s came close to killing off the endeavour, but a fresh discovery of gems in the mid-1940s reignited opal fever.

By the 1960s, Coober Pedy had been recognised as an official town, and consequently obtained both a local government council and a memorable reputation: the 'opal capital of the world'. As much as 70 per cent of the world's total opal production is mined from the earth here (up to 85 per cent if you include the produce from various smaller neighbouring towns). For this,

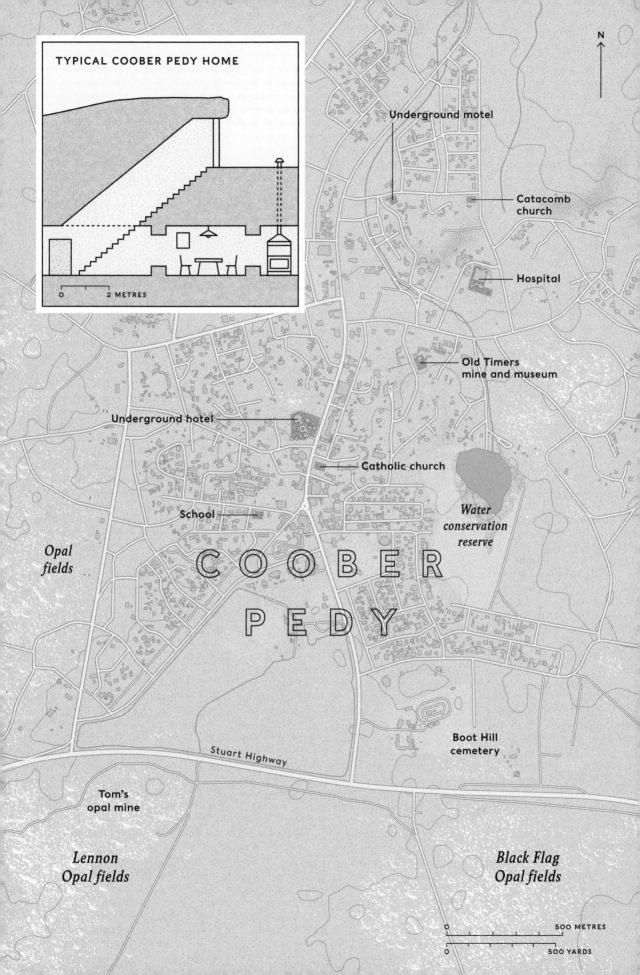

TYPICAL COOBER PEDY HOME

0   2 METRES

N

Underground motel

Catacomb church

Hospital

Old Timers mine and museum

Underground hotel

Catholic church

School

Water conservation reserve

Opal fields

C O O B E R

P E D Y

Stuart Highway

Boot Hill cemetery

Tom's opal mine

Lennon Opal fields

Black Flag Opal fields

0   500 METRES

0   500 YARDS

↑

The landscape around
Coober Pedy is peppered
with punctures in the earth
that are small opal mines.

residents can thank the oceanic waters that flooded this land 150 million years ago. When the sea eventually receded, it left behind residues of hydrated silica hidden in small cracks and fissures in the earth, which, over time, hardened into the valuable stones visible today.

There is just one problem with attempting to build a life in the town of Coober Pedy: the extreme temperatures that ravage the town in summer and cause the mercury to surge to above 50°C (122°F) for up to three or four months at a time, even sometimes creeping towards 60°C (140°F). At such extremes, heat exhaustion and even heat strokes are not unlikely outcomes.

The residents of Coober Pedy came up with an ingenious solution: to head underground. Not just when mining, but for almost the whole of their lives. Many of the original opal prospectors were Australian veterans of trench-style conflict in France and Turkey during the First World War, and so were experienced in constructing underground bunkers. Thousands of people turned subterranean, constructing homes, churches and other necessities in domed caves inside the sandstone rock — often including fake windows to give the illusion of being above ground. An entire town was constructed, visible from the surface as only a scattering of holes that resemble giant anthills, or perhaps a peppering of huge bullet holes.

The contemporary burrows of Coober Pedy, presently home to up to 3,000 people, now include a number of underground hotels, apartments, B&Bs and even underground campsites, to contend with the growing numbers of tourists that this bizarre part of the world has come to attract. Museums, a casino, a pub and a gift shop complete the tourist experience. While the modern luxury of air conditioning has enabled people to withstand the worst of the summer heatwaves more conventionally, allowing the construction of a number of above-ground structures (plus the obligatory grassless golf course), at least half the population — many of whom continue the vocation passed down from their ancestors, taking to the mines in search of ghostly rainbow opals — maintain a traditional bunker lifestyle, living up to 15 metres

(50 feet) below the surface. The rough cost to pay for a tunnelling machine to dig out a five-room underground house is only AU$25,000, making such homes affordable to working people. Furthermore, many families have purchased adjoining homes to create underground 'mansions', living in relative luxury in tunnels that keep them comfortably cool during the summer and warm during the coldest winter nights.

All this means that the residents of Coober Pedy, having dealt with unimaginable heat for decades, are perhaps better structurally equipped than most Australians for the increasingly frequent extreme temperatures forecast to hit the continent over the coming century. Nevertheless, life in such a precarious part of the world is never easy. Concerns have been raised in recent years about the prospect of oil and gas mining in the Arckaringa Basin (to the east of the town) contaminating a local aquifer, the Great Artesian Basin, and thereby potentially threatening Coober Pedy's water supply. Man's desire to strip the Earth of its natural resources may have been responsible for Coober Pedy's birth, but it may also be responsible for its demise.

Everything from homes and hotels to places of worship have been erected below ground.
↓

Cu Chi Tunnels

A SPRAWLING NETWORK OF HAND-BUILT TUNNELS CENTRAL TO THE VIETNAM WAR

VIETNAM

N 11° 04' 19"
E 106° 29' 46"

Le Van Lang and his family ran for their lives. Returning from a military training camp in North Vietnam, he was briefly visiting his home village when they heard the terrifying sound of approaching tanks. The Americans were here, on a search-and-destroy mission. The family fled. Luckily for them, Le Van Lang knew of an asset they had at their disposal. Previously he had constructed a hideout nearby, a secret bunker where the family could take refuge and shelter from the enemy. Despite taking heavy fire from soldiers across the river from their location, this subterranean hideout enabled Le Van Lang and a hastily assembled squadron to shoot back, firing rifles and hurling grenades, before making their escape. Ultimately, the bunker saved their lives.

This was a story played out in some form thousands of times across the bombed-out wasteland of South Vietnam, and was perhaps even the most crucial factor in deciding the outcome of the entire Vietnam War. During the conflict, from 1955 to 1975, the district of Cu Chi, 20 kilometres (12 miles) north of Saigon — modern-day Ho Chi Minh City — became a target under siege in a way few places have experienced before or since. The district was strategically significant to winning the war for both sides, and as such it suffered unimaginable horrors. The steep, forested landscape was relentlessly assaulted day and night with all manner of brutal methods of indiscriminate killing. US pilots were given free rein (indeed, encouraged) to empty their bombs and napalm over the district, all the better to give demolished Cu Chi another show of force than to return to base with unused ammunition.

Under such intense bombardment, the residents of Cu Chi — opposition Viet Cong fighters and collateral-damage villagers alike — did the logical thing, and went to ground. Aiding their cause was an existing network of small hand-made tunnels dug over twenty years earlier, during the country's bloody war of independence from France. As the scale of the Americans' ambitions in trying to crush Communist forces in North Vietnam (and their allies in the South) became ever more apparent, these tunnels once again became a hugely significant piece of infrastructure.

Throughout the 1960s, this tunnel network rapidly grew in size, giving guerrilla fighters major advantages in conducting messy jungle warfare. At their peak, Cu Chi was riddled with hundreds of kilometres of tunnels running beneath the terrain, snaking all the way from the city of Saigon itself to the Cambodian border up to 120 kilometres (75 miles) away. The American troops numbered half a million at their peak, but were consistently on the back foot, blinded by their lack of knowledge about these covert passageways.

The tunnels themselves were a masterclass in camouflage and obfuscation. Their diminutive size — no more than a metre (around 3 feet) wide, by a metre and a half (5 feet) high — enabled them to run beneath the landscape undetected. A network of tiny, almost invisible trapdoors, easily camouflaged with leaves, soil and plants, made the entrances almost impossible to find, let alone penetrate, without prior knowledge of their presence. Furthermore, surprisingly robust engineering tricks such as bamboo-pole reinforcements made them sturdy enough to protect the life of a sheltering guerrilla fighter, often even in the event of a direct hit from a passing bomber.

Despite the heat, darkness, lack of oxygen and frequent presence of snakes, this relative security meant that the tunnels weren't just a hiding place for battalions, but were safe havens for entire villages. Over 20,000 people harboured in these warrens during

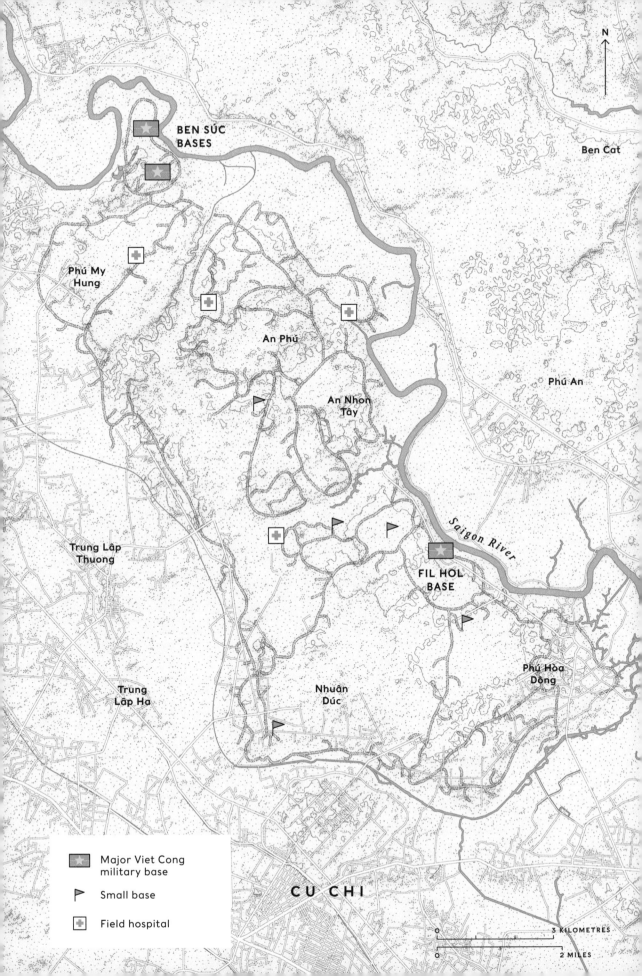

N

BEN SÚC
BASES

Ben Cat

Phú My
Hung

An Phú

An Nhon
Tây

Phú An

Trung Lâp
Thuong

Saigon River

FIL HOL
BASE

Trung
Lâp Ha

Nhuân
Dúc

Phú Hòa
Dông

Major Viet Cong
military base

Small base

Field hospital

CU CHI

0   3 KILOMETRES

0   2 MILES

the course of the war, and they were home to operational hospitals, factories and stables, their occupants kept safe throughout a wave of American-led assaults on the province.

In response to such resilient 'human moles', as US General William Westmoreland famously called the tunnel-dwellers, the American military trained a team of so-called 'tunnel rats' to try to conquer the subterranean world as well as the skies. These specially trained individuals were required to locate, explore and deactivate booby-trapped passageways. But they never even got

← *Small, easily hidden entrances can only hint at the vast warren of tunnels carved across the landscape of Cu Chi.*

close to overcoming the scale of the challenge. With such secure, protective bases from which to launch attacks, the Viet Cong made it almost impossible for even the might of the US military to take control of South Vietnam, let alone defeat their rivals in the North. This tenacity allowed them to simply wait until the US eventually decided to withdraw, before going to take the South in the famous 'Fall of Saigon' in 1975.

An estimated 45,000 Vietnamese died defending the tunnels during the course of the war. Today, most have been destroyed, but those that weren't have become a popular tourist attraction, complete with a rifle range, films depicting archive footage from the bombardment, and a gift shop — as well as specially widened tunnels to accommodate less slender tourists.

Sadly, this isn't the only legacy of the Vietnam War. The widespread planting of mines and booby traps during Vietnam's two-decade conflict (not to mention the aircraft bombs and shells that failed to detonate) means these brutal weapons continue to maim and kill unsuspecting farmers and other rural citizens, with over 100,000 casualties recorded since the war ended. In Vietnam, as with many former war zones, subterranea has a very long memory.

Camp Century

ABANDONED BY THE US MILITARY, CAMP CENTURY'S SECRETS ARE BEING REVEALED BY CLIMATE CHANGE

GREENLAND

N 77° 10' 01"
W 61° 08' 02"

'We have come to the conclusion that experiments with nuclear reactors in Greenland will give rise to a number of problems that we rather … would like to avoid.'

These bland yet firm words were among those submitted in a 1958 memo written by Axel Serup, representative of the Danish Ministry of Foreign Affairs, in response to American intentions to build a revolutionary new scientific research base in Greenland, a Danish territory. The Americans intended to build this base not on the glacial surface, but inside the ice sheet itself, kept heated by an experimental portable nuclear generator. And while a 1951 agreement gave the United States permission to build military bases there for the purpose of defending the island — as indeed it had already done, in large numbers — a nuclear-powered scientific research station entirely within the ice was not something the Danes had anticipated.

Firstly, Serup and colleagues were concerned about negative domestic public reactions regarding nuclear materials being anywhere near Greenland, and secondly — and perhaps most pressingly — the response from their intimidating Soviet rivals in Moscow, who might well react aggressively to the idea of 'nukes' in any form being kept in the territory. With all these concerns, Serup concluded that, frankly, it would be far better if the base were built somewhere else, or perhaps simply not built at all.

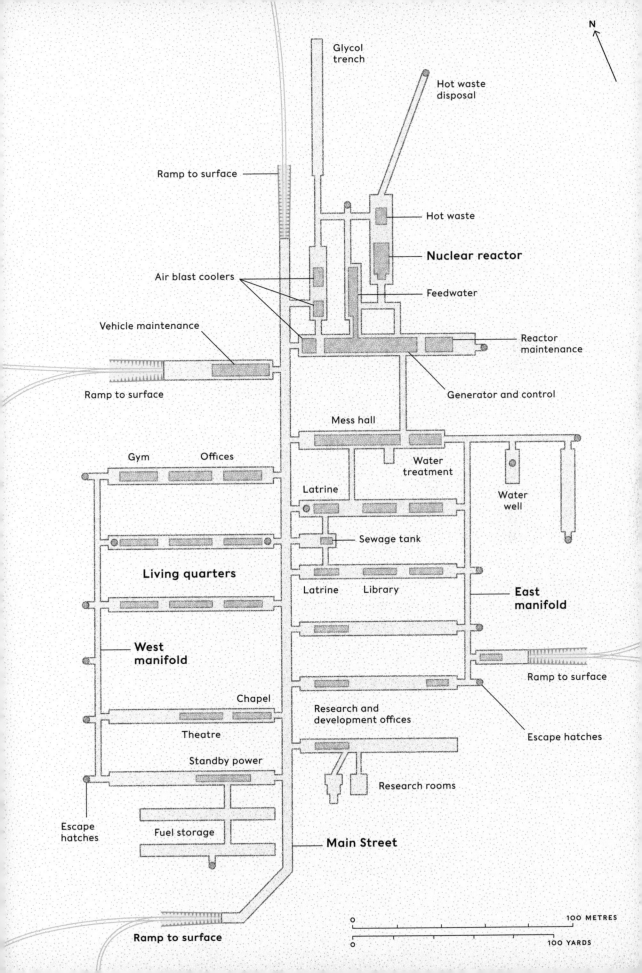

Glycol
trench

Hot waste
disposal

Ramp to surface

Hot waste

Nuclear reactor

Air blast coolers

Feedwater

Vehicle maintenance

Reactor
maintenance

Ramp to surface

Generator and control

Mess hall

Gym   Offices

Water
treatment

Latrine

Water
well

Living quarters

Sewage tank

**East
manifold**

Latrine   Library

**West
manifold**

Chapel

Research and
development offices

Theatre

Ramp to surface

Standby power

Escape hatches

Research rooms

Escape
hatches

Fuel storage

Main Street

Ramp to surface

N

0   100 METRES

0   100 YARDS

Instead, the US pressed ahead, and within a year the construction of the base had gone public, blanketing the media airwaves. The Danes found themselves left with no option other than to pretend they had agreed to the project all along. And it was, initially, a success. Revolutionary construction techniques enabled underground trenches to be quickly and cleanly carved through deep ice sheets. These were then filled with entire buildings, in addition to an underground railway network, a hospital, kitchens, accommodation and all the labs necessary to support up to 200 scientists stationed in this Arctic research centre, situated only 1,300 kilometres (800 miles) from the North Pole. Once covered by curved roofs, these subterranean living and working spaces were quickly hidden by snowfall, becoming invisible to the outside world. Camp Century, as it was dubbed, became the famous 'city under ice', covering an impressive 2,500 square metres (3,000 square yards). For nearly a decade, it successfully gave the ever-ambitious US a foothold in the Arctic Circle.

Yet the logistical doubters had their victory as well. Eventually, the immense force of the ice — moving faster than expected — began to destabilise the walls and infrastructure of the site, threatening the integrity of the whole enterprise. With these concerns in mind, closure of the base began during the mid-1960s. All staff were removed and, with minimal decommissioning, Camp Century was finally shut down in 1967. The facility was deserted, left to the whims of the harsh conditions of the Arctic. When a team was sent to survey the site in 1969, they found the large walkways had been reduced to crawl tunnels, and the supporting metal arches had been twisted and deformed beyond recognition. The unstoppable power of the ice sheet had crushed the life out of Camp Century.

Over half a century later, it might be expected that, as intended, there would be few remaining traces of such a facility visible. That over fifty years of falling snow would have buried the railway and buildings until the end of time. And yes, the site is now 30 metres (roughly 100 feet) below the surface of the ice. But since 2017, this location has also been home to the Camp Century

The construction of a US research base in the Greenland ice sheet in the 1950s was considered a monumental triumph of human ingenuity over nature.

→

Climate Monitoring Programme. Their job: to collect vital data — temperatures, wind speeds, humidity and atmospheric pressure — and to measure and track the changing ice conditions at Camp Century. As that moving ice perhaps indicated, climate change through the course of the twentieth century is now rapidly accelerating the melting of the ice sheets under which Camp Century is buried. The concern is that this buried debris is on its way back to the surface.

And there's a twist. The Kremlin would have been right to be suspicious. Camp Century wasn't simply the scientific research station it was claimed to be, but instead, perhaps predictably, a military base. Known as Project Iceworm, the base gave the US a strategic advantage during the Cold War. Equipped with nuclear warheads, it was designed to be a launching site capable of hitting the Soviet Union with hundreds of intercontinental ballistic missiles. The reaction chamber of the nuclear generator was removed in 1964, but everything else, including thousands of tonnes of radioactive waste and toxic chemicals, was left to the elements, to be entombed in snow for, supposedly, eternity. Yet by the end of the century, the site is expected to be experiencing more ice melt than snow accumulation, reversing this process.

The increasingly likely re-emergence of this hazardous detritus is of serious concern to Greenland authorities. Meltwater pouring out from underneath the glaciers, having run through where Camp Century is buried, is already potentially contaminated by chemicals from the facility, with diesel fuel, radioactive coolants and persistent organic pollutants (POPs) such as polychlorinated biphenyls (PCBs) now entering the marine environment. Such pollutants were supposed to remain entombed for ever, yet this short-term thinking has now been exposed.

Crucially, Camp Century is not alone. While it might be the most high profile, and perhaps also the most high risk, there were four other bases built into the ice sheet in the same way during this time. Furthermore, as many as thirty bases were abandoned by the US military in Greenland during the last century. One such example is Bluie East Two, a former airfield, now laden with thousands of rusting diesel drums, asbestos-filled buildings, and likely consignments of dynamite. While not all subterranean, many of these sites experienced a similar fate to Camp Century, and so each is likely to be undergoing a similar experience, slowly decomposing into the environment, along with all the unpleasant materials contained within. Similar to how melting permafrost across Siberia is allowing deadly pathogens such as anthrax to re-emerge, infect and kill both livestock and people, a changing climate is enabling once-forgotten hazards to become major threats once again.

Great Man-Made River

IMMENSE PIPELINES CARRYING VITAL
WATER SUPPLIES UNDERNEATH HUNDREDS
OF MILES OF DESERT

LIBYA

N 25° 55' 19"
E 17° 23' 17"

The year was 1983. Fourteen years since revolution had swept aside the old order in Libya and installed Muammar Gaddafi as leader of the nation, a major crisis was building. Fresh water was becoming an increasingly scarce resource, as traditional river flows were insufficient to supply an expanding agricultural sector and a growing population in the cities of Tripoli, Benghazi and other urban areas. Furthermore, encroaching seawater threatened to contaminate the precariously placed rain-filled aquifers along the North African nation's Mediterranean coastline. Decisive, ambitious action was needed. And it came, in the form of what's been called the world's biggest irrigation project.

Ten thousand years ago, at the end of the last ice age, North Africa was a lush forest. Over millennia, a gradually drying climate turned this temperate environment into the largest sand desert on Earth, hence the name: Sahara (translated from the Arabic ṣaḥrā, meaning 'desert'). But below ground, the remnants of this ancient world still exist, in the form of huge groundwater reserves. Studies have shown Africa to be home to an estimated 660,000 cubic kilometres of groundwater — larger than the volume of the Black Sea, and enough to fill the Grand Canyon over 150 times. Crucially, it's more than a hundred times what is currently found on the continent's surface. This isn't just below obviously wetter regions such as the Congo rainforest or Okavango Delta — the mighty Sahara is in fact home to the largest identified reservoirs of groundwater in the entire continent.

Consequently, this meant that when oil exploration began in the 1950s, the Libyans were amazed to discover part of what has since become known as the Nubian Sandstone Aquifer System — the largest fossilised water aquifer in the world (also stretching underneath neighbouring Chad, Egypt and Sudan). It is estimated to contain as much as 150,000 cubic kilometres, equivalent to nearly twice the volume of the Caspian Sea. Three decades later, this aquifer was seen as the solution to the country's growing water crisis. The only problem: tapping into these vast reserves meant the nation's citizens being reliant on water being safely transported across hundreds of miles of open desert, a challenge that led to undoubtedly one of the largest civil engineering projects ever undertaken.

The government decided to pursue one of their cheapest options (although still costing tens of billions of dollars) to create what has become known as the Great Man-Made River (GMR) — an artificial network of subterranean pipelines that would simultaneously solve the water problem and showcase the modern Libyan nation's ingenuity and engineering prowess. Indeed, Gaddafi and the government frequently referred to it as the 'eighth wonder of the world'. The first foundations were laid in 1984, and saw a lengthy tunnel burrowed deep into the southeast of the country, all the way to the aquifers drilled in Tāzirbū and Sarīr. Four-metre-diameter (13-feet) pipelines were laid in trenches dug 7 metres (23 feet) down into the desert, then buried beneath the sand by bulldozers.

The outcome was transformative. Hundreds of thousands, if not millions, had their lives significantly improved once the first of five planned stages opened in 1991. Dramatic photos show huge crowds assembling in fast-flowing waters to celebrate the arrival of this fossil water in Benghazi, where saltwater intrusion was hitting most severely. The second phase, completed in 1996, expanded the reach of the GMR to include the capital, Tripoli, now encompassing as much as 70 per cent of the country's

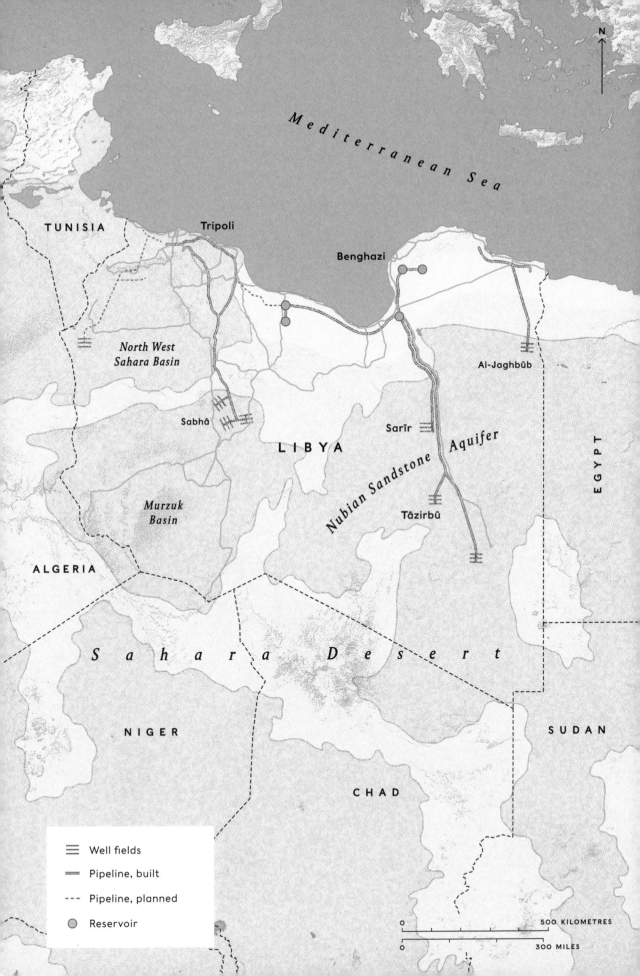

N

Mediterranean Sea

TUNISIA

Tripoli

Benghazi

North West
Sahara Basin

Al-Jaghbūb

Sabhā

Sarīr

LIBYA

EGYPT

Murzuk
Basin

Nubian Sandstone Aquifer

Tāzirbū

ALGERIA

Sahara Desert

NIGER

SUDAN

CHAD

≡≡≡ Well fields

▬▬▬ Pipeline, built

- - - Pipeline, planned

● Reservoir

0   500 KILOMETRES

0   300 MILES

population. While the entire network has yet to be fully constructed, a complete GMR would consist of around 4,000 kilometres (2,500 miles) of pipeline, capable of pumping 6.5 million cubic metres (230 million cubic feet) of water per day.

It might not be obvious from surface level, but this underground network has arguably kept Libya alive. Even in the face of the extreme violence that erupted during the 2010s, the continued functioning of the GMR, submerged below the desert, has kept the country on life support. When a section of the network was bombed in 2011, it led to a heated war of words between government forces and the intervening NATO military, each accusing the other of risking the lives of millions of civilians by interrupting the vital flow from this mighty piece of underground infrastructure. While the longevity of the steel-reinforced concrete and other materials used to build the GMR remains in doubt (the oft-quoted fifty-year lifespan suggests an expiry date is rapidly approaching), the enormous quantities of water in the country's aquifers mean Libya should be able to enjoy a glut of fresh water for hundreds or even thousands of years to come.

A completed network would require installing 4,000 kilometres (2,500 miles) of pipeline, costing billions of dollars.
↓

Sonnenberg

BUILT TO FULFIL NATIONAL LEGISLATION
DEMANDING SHELTER FOR THE ENTIRE
POPULATION

SWITZERLAND

N 47° 02' 47"
E 08° 17' 42"

Every day, tens of thousands of motorists — often transiting between Germany and Italy — drive beneath Sonnenberg mountain, along a pair of unassuming mile-long motorway tunnels that lead into the Swiss city of Lucerne. Grey and spartan, the tunnels are dark and, to the untrained eye, appear to have very little remarkable to say about themselves. But perhaps a few of these passing drivers will know the full story of this extraordinary piece of engineering, and the revelation that this tunnel was also once the largest civilian bunker in the world.

Perhaps primarily most famous for chalets, chocolate, watches and global finance, Switzerland also happens to be one of the world leaders in nuclear bunkers. The landlocked Alpine nation passed a law in 1963 stating that every citizen must have access to a safe bunker in the event of a nuclear conflict. These could be private shelters, built into homes, or large community shelters. The fear was that the Cold War might turn hot, and that radiation from nuclear fallout would become a likely component of the national future. The outcome: bunkers spread liberally across Switzerland, peppering the landscape with holes like the cheese the nation is renowned for. By the turn of the twenty-first century, there were as many as 300,000 private bunkers spread across Switzerland, in addition to 5,000 public bunkers, sufficient to shelter the entire population.

Sonnenberg dates back to October 1976, when two large road tunnels were opened to the general public, with a 40 million Swiss franc price tag. While they may have appeared entirely ordinary, in the event of an atomic bomb warning, they were built to be capable of quickly converting into a secure shelter, with bunk beds for up to 20,000 people (around a quarter of the population of above-ground Lucerne, including a staff workforce of 700). In the middle sat a seven-storey subterranean nerve centre, from where the logistical operation of this huge facility would have taken place, including hospital amenities, a radio studio and a small prison. All this was expensive. Even just keeping the hospital on standby cost an incredible 250,000 francs per year, simply to maintain a supply of medicines and operational equipment in case the emergency scenario ever unfolded.

Sonnenberg's secondary purpose has never been utilised, and thankfully so, for a long list of reasons. Firstly, an estimated two-week turnaround time to prepare the facility for usage, such as unfolding the thousands of flatpacked bunk beds, pours cold water on the idea that the residents would have been able to rely on the shelter for their immediate survival (nuclear strikes tend not to provide such an extended notice period). Furthermore, had the facility actually been ready in time, it is rumoured that the shelter's 350,000-kilogram concrete blast doors, each one and a half metres thick, would have failed to adequately protect those inside from the nuclear fallout from which they were supposed to be sheltering. Even if the doors had worked, assuming the facility was at capacity, there would have been a severe shortage of portable toilets and functional showers. Finally, frenzied citizens were also requested to bring a two-week supply of food, with the authorities deeming it impossible to adequately feed so many people themselves. After a fortnight, the water supply was expected to run out, and so everyone would be sent back to the surface to face whatever horrors might await them.

Thankfully for the residents of Lucerne — and indeed the rest of the world — the subsequent years have seen the threat

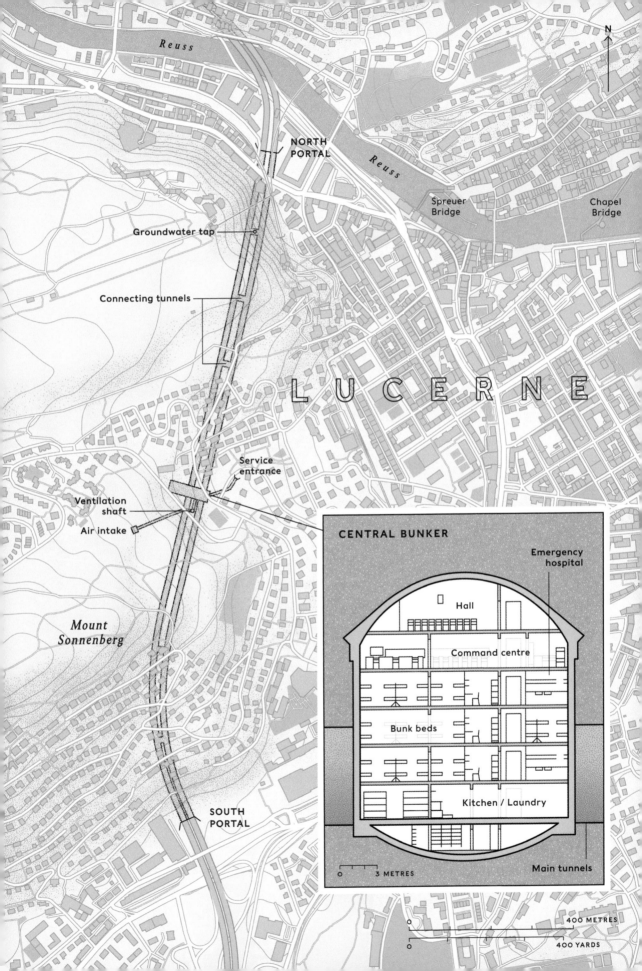

N

Reuss

NORTH
PORTAL

Reuss

Spreuer
Bridge

Chapel
Bridge

Groundwater tap

Connecting tunnels

L U C E R N E

Service
entrance

Ventilation
shaft

Air intake

*Mount
Sonnenberg*

SOUTH
PORTAL

CENTRAL BUNKER

Emergency
hospital

Hall

Command centre

Bunk beds

Kitchen / Laundry

Main tunnels

0   3 METRES

0   400 METRES

0   400 YARDS

→

Sonnenberg's basic facilities claimed to be able to host 20,000 local residents in the event of a nuclear attack.

of nuclear war gradually recede. While the bunker law hasn't been repealed, the anticipated threats in which they might be necessary are now more likely to include natural disasters or devastating global pandemics than nuclear warfare. Sonnenberg — scaled down in 2006 to hold just 2,000 people — might today simply be a glorified tunnel. However, many other bunkers around the country have ended up with far more interesting post-nuclear lives.

For example, Sasso San Gottardo, a former Second World War fort built into the side of a Swiss mountain, now operates as a subterranean museum, documenting how it functioned during wartime. Hotel la Claustra is a former maze of shelter tunnels now turned into a luxury hotel, while Seiler Käserei is a large-scale cheese manufacturer operating out of another former bunker. Many of Switzerland's thousands of unused shelters have also been suggested as feasible emergency housing for homeless people or refugees, while some larger spaces have reportedly been snapped up for use as data centres, or doomsday safe havens for the ultra wealthy.

Darvaza Gas Crater

THE FABLED 'DOOR TO HELL', ACCIDENTALLY CREATED BY SOVIET GAS EXPLORERS

TURKMENISTAN

N 40° 15′ 08″
E 58° 26′ 16″

'Abandon all hope, ye who enter here.' These oft-quoted words (or some version of them, the original line having been written in author Dante Alighieri's native Italian) are, according to the fourteenth-century poem *Inferno*, among those printed at the entrance of the door leading to Hell. Yet while Dante may have painted the vividly harrowing picture of the afterlife that endures in popular imagination today, he gave no clue as to where exactly in the world such an entrance is to be found.

In fact there is a plethora of so-called 'gates to hell' spread across the surface of the planet. All feature some eye-catching, often fiery, characteristics that have captured local imaginations and led to dramatic stories about flaming paths to a punishing underworld of pain and torture — just as Dante envisaged seven centuries ago. These locations include Mount Etna, whose spectacular eruptions on the island of Sicily saw it gain the medieval reputation of a likely route to Hell. In Iceland, the destructive ash clouds, lava flows and flaming pumice that sporadically burst forth from the active Hekla volcano saw it regarded locally as an entrance to the abyss. And in a smoking, sulphurous region of northern Japan, folklore depicts the caldera of Mount Osore as another route to the netherworld — the mythological Sanzu River being a symbolic threshold the deceased must cross in order to reach the afterlife.

Yet undoubtedly among the most enigmatic of these 'gates' is the Darvaza gas crater, Turkmenistan's own entrance to Hell

K A R A K U M

D E S E R T

Darvaza
Gas Crater

Darvaza

0   100 METRES

0   100 YARDS

(indeed, *darvaza* is a corruption of Farsi for 'the gate'). Situated roughly 250 kilometres (155 miles) from Ashgabat, the Turkmen capital, it sits alone amid the silent expanse of the Karakum Desert, a bowl-like crater stretching roughly 69 metres (226 feet) across and 30 metres (98 feet) down at the deepest point. If the commonly told story is to be believed, it's a striking lesson in hubris and humility in the face of unanswered questions about the hidden subterranean world.

Legend tells that, in 1971, Soviet miners were drilling in this region when the ground began to crumble, and an enormous cavern suddenly opened beneath their feet. Having established that this spontaneous crater (and a few smaller ones that simultaneously opened nearby) was burping out large quantities of toxic fumes such as methane, the decision was made to set the gas alight, then wait a few days or perhaps weeks until it had burnt off, leaving nothing but an empty hole in the ground. Half a century later, the subsequent fire is still fiercely burning away, defying any predictions that the fuel supply will run out any time soon.

Is any of that true? Maybe. Maybe not. There appears to be little concrete evidence one way or the other, merely the ongoing repetition of this least urban of all urban myths. What the seemingly never-ending fires do hint at is the extent of the gas supply contained beneath the surface, and the immense scale of the various subterranean cavities required to contain such a large quantity of gas.

Temperatures of up to a rumoured 1,000°C (1,830°F) successfully deterred any major scientific expeditions into the crater for decades. But in November 2013, Canadian George Kourounis — self-proclaimed 'adventurer' and 'storm chaser' — became the first recorded person to make the bold leap into the depths of Darvaza. To withstand the intensity, he wore a futuristic heat-reflective suit equipped with a specially designed breathing apparatus and heat-resistant harness, allowing him to be lowered into the blazing pit. His hope: to provide an answer to the very simple but valuable question of whether life — even just small microbes — could exist in such extreme conditions.

↑
The fires of Darvaza have been burning for half a century, with no indication that the gas supply is even close to running out.

After successfully surviving both his toasty descent and return to the surface, Kourounis was able to confirm that, yes, he had found bacteria at the bottom of the crater, happily living in the very midst of the inferno. Indeed, it was bacteria not found in the soil surrounding the crater, suggesting a unique ecosystem surviving in the immense scorching temperatures of Darvaza.

As Kourounis has hinted since his endeavour, there are valuable incentives to learning about organisms in the roaring fires of Darvaza as a proxy for potential life elsewhere in the universe. The presence of bacteria underground in such extreme conditions opens the door to discovering microbes (or possibly more developed species) on super-hot planets, perhaps with methane-filled atmospheres, when previously such ideas would have been dismissed as ridiculous. The secrets of subterranea and outer space perhaps have more in common than we might imagine.

G-Cans

THE WORLD'S LARGEST UNDERGROUND FLOOD DIVERSION FACILITY, BUILT TO PROTECT TOKYO

JAPAN

N 35° 59' 32"
E 139° 46' 44"

The mighty floodwater discharge facilities beneath Tokyo have been compared to a cathedral.
→

In 1958, Typhoon Kanogawa smashed its way through downtown Tokyo, Japan. Wind speeds approaching 200 kph (125 mph) unleashed heavy rain that triggered hundreds of landslides and flooded half a million homes. More than 200 people lost their lives in the city, with over 1,200 people perishing across the entire country.

But this wasn't a one-off incident. Instead, it was just one chapter in a brutal century of flooding for Tokyo, starting with the heavy rains that inundated the city in 1910, leaving nearly 200,000 homes flooded and hundreds of people missing. In 1917, it happened again — high tides costing the lives of over a thousand people — and was followed by a succession of powerful and destructive typhoons that swept through in the late 1940s. Various protective measures such as tide barriers and water gates were installed after the devastating events of 1958, yet the latter half of the century saw the city continue to be tormented by seasonal downpours that had nowhere to flow except into the streets and peoples' homes, including one major typhoon in 1966, and another in 1979, when Japan was battered by Typhoon Tip, possibly the strongest storm ever recorded.

If you were deliberately designing a city to be prone to flooding, you'd be hard-pressed to locate it more perfectly than Tokyo. The city was founded on the banks of not just the Edo River (a reminder of the city's pre-1868 name, Edo) but also the Ara, Naka,

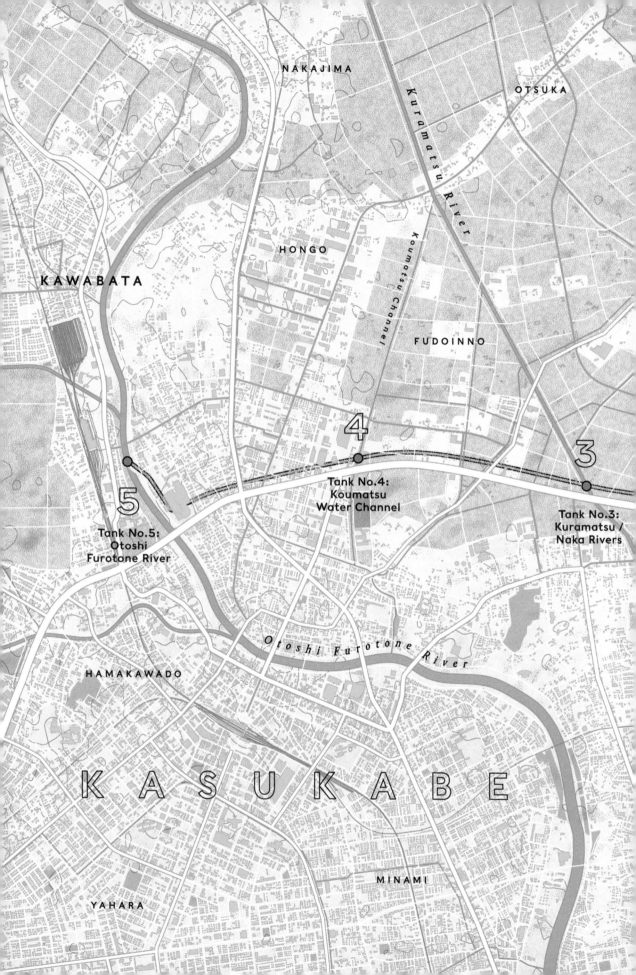

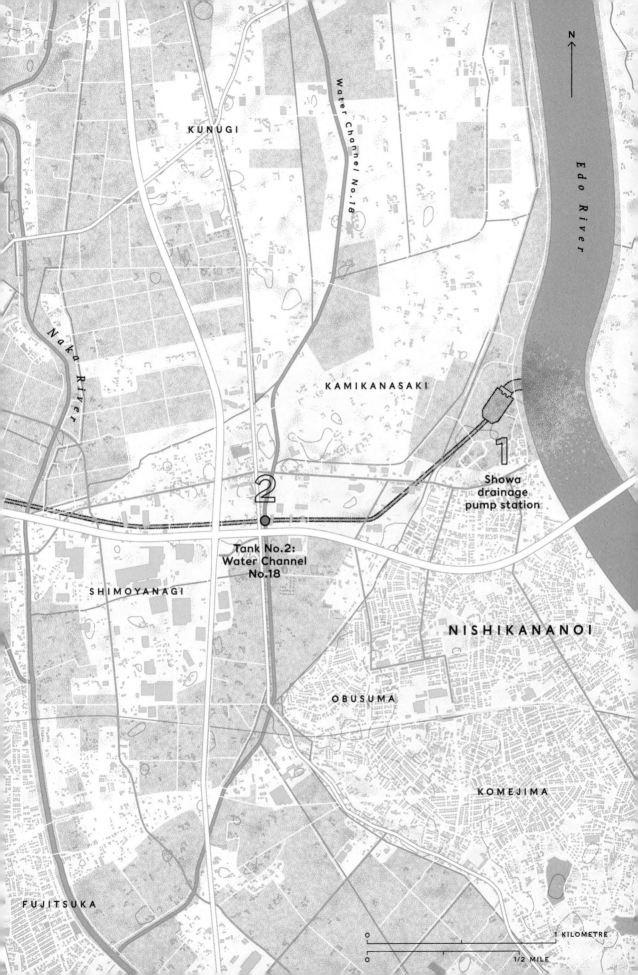

N

KUNUGI

Water Channel No.18

Edo River

Naka River

KAMIKANASAKI

1

Showa
drainage
pump station

2

Tank No.2:
Water Channel
No.18

NISHIKANANOI

SHIMOYANAGI

OBUSUMA

KOMEJIMA

FUJITSUKA

0   1 KILOMETRE

0   1/2 MILE

Ayase and Sumida rivers, with many smaller channels nearby. Being positioned near to so many waterways, while highly advantageous for international trade, increasingly became a hazard as the city grew. The slow-flowing alluvial plains surrounding Tokyo Bay only add to the problem, as water takes a very long time to exit the floodplains into the ocean, leaving no room for further torrents still gushing down from higher ground.

By the end of the 1980s, a decade in which Tokyo had again been subjected to a series of almost annual flooding events — in part due to an ever-growing population living in an urban environment that was spreading outwards, paving over the surrounding marshlands and rice paddy fields — Japanese authorities decided to take significant action. A plan was hatched, fully commissioned in 1992, to begin construction on what was dubbed the 'world's largest drain' — a long-term infrastructure project that would be capable of managing any future flooding events that threatened the safety and prosperity of the city's tens of millions of residents.

When it was finally completed, in the late 2000s, the final product — the Metropolitan Area Outer Underground Discharge Channel, fondly nicknamed the 'G-Cans' project — was something truly colossal. Buried beneath the city streets lies the world's largest floodwater diversion facility, a pressurised system capable of redirecting excess water with ease. The cavernous interior of the main water tank, 22 metres (72 feet) below ground, is intimidatingly vast. Fifty-nine pillars, which appear even larger than their 18-metre (59-feet) height, separate the floor from the ceiling, leaving anyone walking around the base resembling ants walking around the hull of an empty ship. At 177 metres (581 feet) long and 78 metres (256 feet) across, the facility stretches almost out of sight. With majestic infrastructure on such a massive scale, it's no wonder the facility has been compared to a coliseum, or even a cathedral.

Five huge cisterns — each large enough to fit London's Nelson's Column inside — have the pumping power necessary to remove an Olympic swimming pool of water from this storage facility in a matter of seconds. Firing the water through a tunnel network over 6 kilometres (nearly 4 miles) in length, they ultimately dump the unwanted floodwater into the large Edo River, which carries it to the ocean.

At present, such a process is required to take place an average of seven times per year. But the future is somewhat more uncertain. As the city grows, the rains increase and the seas rise, many people in Tokyo are vocally worrying about whether their underground pumping system is large enough to cope with the impact of a more unpredictable climate. The livelihoods of millions, and lives of thousands, are at stake. Authorities are now being forced to confront the uncomfortable possibility that even this huge facility might not be sufficient to contend with future flooding events. Tokyo's twenty-first century might yet be as tumultuous as the last.

As the world's largest underground floodwater diversion facility, Tokyo's G-Cans requires careful coordination from central control.
↓

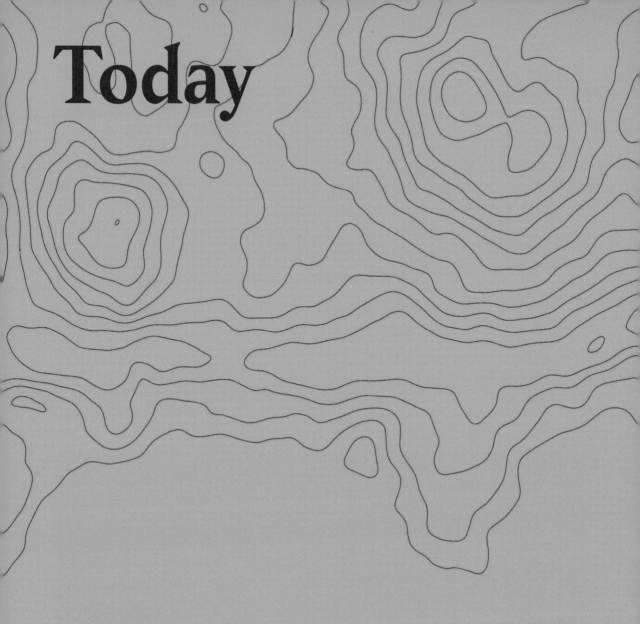

Today

Large Hadron Collider

THE WORLD'S LARGEST AND MOST POWERFUL PARTICLE ACCELERATOR

SWITZERLAND / FRANCE

N 46° 16' 26"
E 06° 04' 19"

'End of the world!' shouted tabloid headlines. The date was September 2008, and journalists had just uncovered a decade-long plot by scientists in Switzerland to build a super-powerful machine capable of undertaking what was quickly nicknamed the 'Big Bang experiment'. This experiment, they explained, would involve smashing subatomic particles together at close to the speed of light, simulating the conditions immediately after the creation of the universe. In a worst-case scenario, these prophets of doom assured their readers, we might witness the formation of small black holes, which would quickly expand to consume Earth and everyone on it. Legal action was even taken to try to prevent the machine being switched on, so great was the hysteria.

The reality was somewhat less apocalyptic, although just as newsworthy. What had actually happened was that the European Organisation for Nuclear Research, the Geneva-based European atomic physics laboratory more commonly known as CERN (responsible for, among other things, the creation of the World Wide Web, a key component of the modern internet), had spent years upgrading their Large Electron-Positron Collider (LEP). The LEP, a gigantic machine commissioned back in May 1981, was constructed for the purpose of intentionally colliding tiny electrons with their antimatter counterparts, positrons, to further the study of bosons — subatomic particles that are essential to the workings of quantum mechanics. This required the construction of a

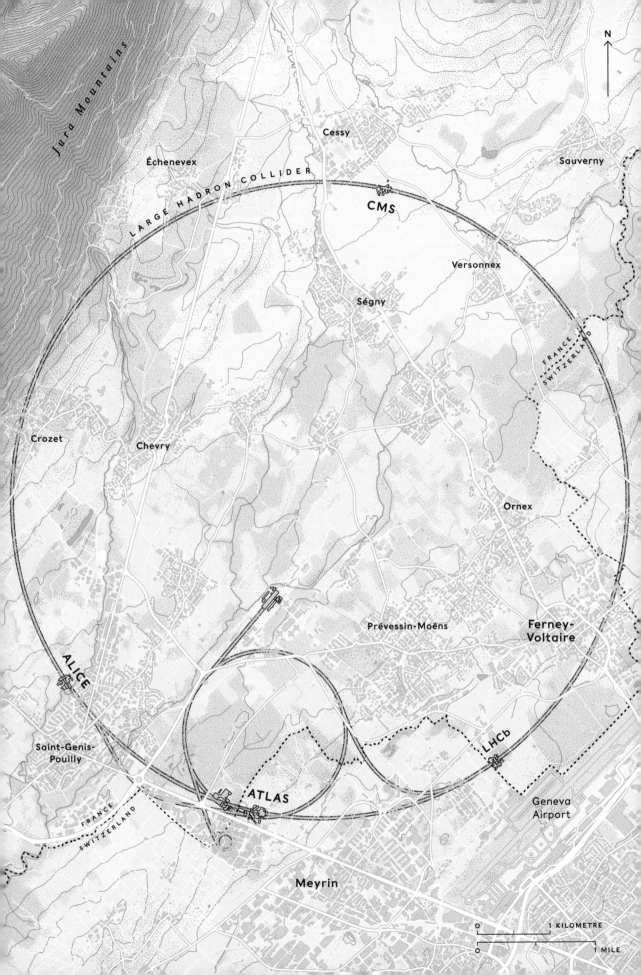

27-kilometre (nearly 17-mile) circular tunnel below Geneva. It took three tunnel-boring machines three years to excavate the area, making it Europe's largest civil engineering project at the time. The tunnel was finally completed in February 1988, with the LEP becoming active the following summer.

The new, ambitious upgrade — installing the world's largest and most powerful particle accelerator within the existing infrastructure of the LEP's tunnel — began as far back as 1984. It promised a leap forward in research, potentially uncovering some of the most profound scientific answers about the nature of the universe. Yet it was questioned whether such a machine was really necessary, especially given the American interest in building their own 'Superconducting Super Collider' in an 87-kilometre (54-mile) tunnel — longer than the entire Panama Canal — in Waxahachie, Texas. Despite objections, proposals for the European project were pushed ahead, a decision that was justified in 1993 when the US government, two years after construction began, voted to cancel their project due to rising costs.

A year later, the Large Hadron Collider (LHC) was officially approved for construction. The goal: to create a machine capable of smashing together high-energy beams of protons travelling at close to the speed of light, unlocking various mysteries about the nature of quarks and other particles emitted during the mini 'fireball' that momentarily ignites in the aftermath of collisions at such immense speeds.

When the power switch was finally flicked on for the LHC, it didn't go entirely to plan. The world didn't end, but initial experiments certainly did, when a fault caused cooling fluid to leak into the tunnel. Thirty-seven magnets needed to be replaced before the LHC was eventually up and running properly. But the scientists' perseverance would be rewarded a few years later, when the fabled Higgs boson was successfully identified in July 2012. The so-called 'God particle' had first been predicted to exist by British physicist Peter Higgs nearly half a century earlier, and subsequently led to him jointly winning the Nobel Prize for Physics the following year.

↑
The futuristic tunnel containing the Large Hadron Collider is as devoid of atmosphere as the moon, and cooled to almost absolute zero.

The LHC's tunnel sits 100 metres (328 feet) beneath the French–Swiss border, with the mundane car parks and office blocks visible on the outskirts of Geneva giving little clue to the bizarre experimentation taking place below ground. When active, the tunnel is almost unfathomably cold. In order to create the very strong magnetic field necessary for the experiments to operate, thousands of superconducting electromagnets line the tunnel, cryogenically chilled with the help of 120 tonnes of liquid helium to -271.3°C (-456°F) — almost at the fabled 'absolute zero'. It is also the biggest operational vacuum system in the world, as empty as the atmosphere found on the moon.

While the LHC is far from retirement age (it's supposed to run for at least twenty years), there have already been official plans announced for a further upgrade for it to become the 'High-Luminosity LHC', supposedly operational from 2026. Increasing the luminosity (a measure of brightness, or radiation) by a factor of ten, this upgrade would enable more collisions to take place, producing millions more bosons to study.

Yet even this contraption might appear something of a minnow compared to the plans released by CERN for a new particle accelerator. If constructed, the 'Future Circular Collider (FCC)' would be built in a new tunnel 100 kilometres (62 miles) long or more — four times the size and six times as powerful as the LHC. The subterranean world beneath Geneva may just be beginning its scientific adventure.

Otay Mesa

DOZENS OF SMUGGLING TUNNELS BETWEEN MEXICO AND THE USA

USA / MEXICO

N 32° 32′ 45″
W 116° 58′ 14″

Above ground, symbolic lines cross continents, separating the Earth's surface into welcoming and unwelcoming spaces. Officers check paperwork, demand answers, grant life-changing decisions. Rangers patrol up and down these borders, driving around in SUVs on the lookout for weaknesses in their defensive infrastructure. But beneath their feet lies a different world. Defying the binary framework enforced at surface level, down here there is both the motive and the ability to break the rules. Down here, boundaries can be ignored, surveillance is blind, paperwork is irrelevant, and the lines delineating territorial sovereignty become far more blurred.

In contemporary times, it's hard for geology not to become political. Land has immense symbolic meaning, especially when overlaid with cultural narratives that are underpinned by one political doctrine or another. Where nationalism reigns, borders become hotspots of tension, fear, resentment and, often, conflict. While this doesn't always extend to outright violence, low-level discord can arise through the simple restrictions placed on what (and who) is and isn't allowed to move across these often arbitrary lines. As they say, however, rules are made for breaking. And when borders become a persistent problem, subterranea offers an exciting new dimension in which to try to circumvent those rules.

One of the most closely observed and highly politicised national borders in the world — as well as one of the most heavily fortified — runs 3,111 km (1,933 miles) from the Pacific beaches of Tijuana to the mouth of the Rio Grande as it pours into the Gulf of Mexico, and separates the United States of America from the United Mexican States. A journey along this line — a third of which is fenced — would bring one in close contact with everything from rugged mountains to winding rivers to shifting desert sands. Preventing people (and certain banned substances) from passing across this terrain has become an American obsession. But there is one area where the underground world offers a way past the intense scrutiny of the surface.

Millions of years ago, Otay Mesa was a marine terrace deep underwater. As the eastward San Ysidro Mountains rose, a topsoil was formed across the region, deposited by water running off the mountains and leaving behind a low-level plateau of sandstone, silts and clays. This short strip now sits on the border between northern Mexico and southern California — more specifically, between Tijuana airport and the Otay Mesa neighbourhood of San Diego. And the unique geology of this semi-arid spot makes it perfect for one particular activity: tunnelling. By avoiding the soggy, sandy soils to the west, and the impenetrably hard volcanic rock of the San Ysidro Mountains to the east, warrens can be cut down here through the malleable but firm bentonite clay — which often requires little more than a shovel, and almost no external supports — underneath the border, and up the other side.

Would-be drug smugglers (especially the organised crime syndicate known as the Sinaloa Cartel, masterminded in the late 1980s and early 90s by the infamous Joaquín 'El Chapo' Guzmán) have thereby been able to successfully transport their illicit goods, such as marijuana grown in the Mexican states of Sinaloa and Sonora, to the lucrative commercial opportunities offered by the US. Such tunnels also make it easier for human smugglers to try to slip aspiring migrants under the borders unnoticed. Most helpfully, the busy industry surrounding the area where San Diego and Tijuana rub up against each other means that the heavy-duty tools and equipment required for this enterprise don't stand out as unusual or suspicious — especially when the entrances are

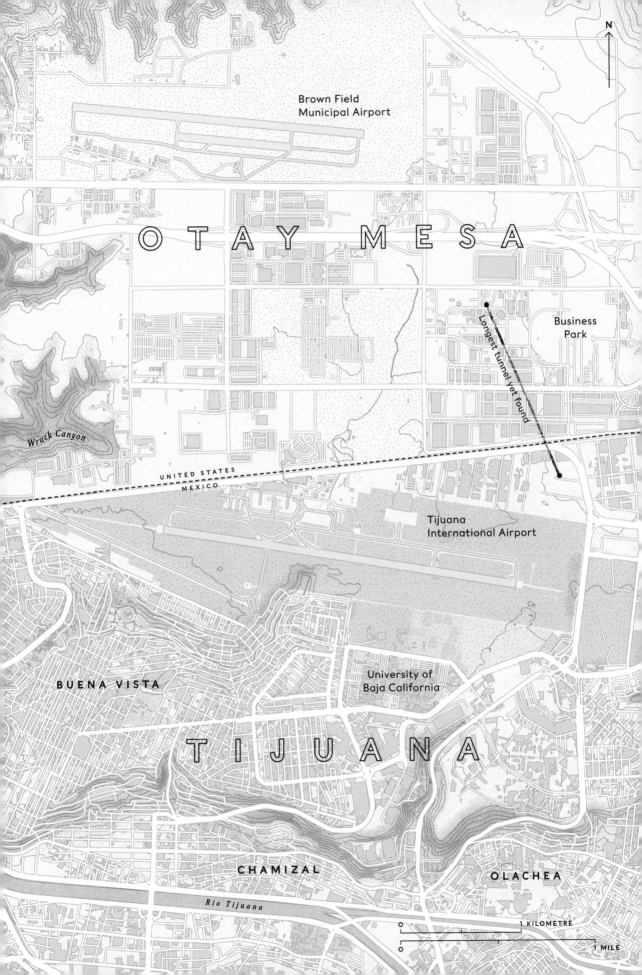

N

Brown Field
Municipal Airport

O T A Y   M E S A

Business
Park

Longest tunnel yet found

Wruck Canyon

UNITED STATES
MEXICO

Tijuana
International Airport

University of
Baja California

BUENA VISTA

T I J U A N A

CHAMIZAL

OLACHEA

Rio Tijuana

0   1 KILOMETRE

0   1 MILE

hidden within quiet commercial premises, including warehouses and industrial-sized waste facilities. There have been hundreds if not thousands of illegal tunnels constructed beneath the US–Mexico border throughout history, but few locations offer such perfect conditions for it as Otay Mesa.

One tunnel discovered in the area in August 2019 is a prime example of how remarkably sophisticated these engineering projects have become. At 1,313 metres (4,309 feet) in total, it was the longest recorded tunnel yet found running through the Otay Mesa area, successfully crossing underneath the border at an average 21 metres (70 feet) below the surface, and tall enough for an adult to stand up in. It was found fully equipped with a railway line running the entire length, as well as lighting, functioning ventilation, high-voltage electrical cables and a system to pump out unwanted water. Other tunnels have been discovered that even replace humble ladders at either end with working elevators, capable of moving large quantities of narcotics through the tunnel at great speed.

Once located by authorities, these tunnels tend to be quickly eradicated, filled in and wiped from existence, to prevent opportunistic usage by prospective future miscreants. Yet while border agents might be wise to the tactics their tunnel-dwelling rivals are deploying, effective detection remains elusive. The scale of operations in the Otay Mesa area and the ever-evolving tricks adopted by these professional criminals — including tunnels dug into the basements of rented houses, or trucks with removable floors designed to transport goods while parked atop open manholes — mean this contraband arms race looks set to continue to escalate (somewhat counter-metaphorically) deeper and deeper into the earth.

In Otay Mesa, criminals build long, sophisticated tunnels to smuggle contraband past the strict
← *authorities above ground.*

Guatemala City Sinkholes

WHERE HUGE SINKHOLES SUDDENLY SWALLOW CITY STREETS

GUATEMALA

N 14° 36′ 49″
W 90° 30′ 42″

Strong winds howling, heavy rains lashing, streets turned into rivers. As tropical storm Agatha made landfall from the Pacific in late May 2010, it left a trail of devastation in its wake, maiming and killing hundreds across Central America. Now it was bearing down on the capital, Guatemala City. As the city's million or so residents sought shelter to wait out the start of the annual hurricane season, one building in Zone 2 unmistakably felt the earth shift. Sitting on the corner of Avenue 11A and Street 6A in a quiet northern suburb, this three-storey textile factory suddenly, without warning, collapsed into the ground. In a matter of seconds, the foundations had completely given way, and the entire building was gone, vanishing into a huge hole. Shocked neighbours found themselves teetering on the edge of a precipice that appeared, to all intents and purposes, to lead into the centre of the Earth.

Guatemala City had just been hit by a sinkhole, and a big one as well. Sinkholes (also sometimes called dolines) occur naturally when surface levels collapse, exposing empty spaces that have been surreptitiously created below ground. This is often caused by water, either eroding the supporting rock through the action of underground rivers and streams, or simply saturating the surface soils to the point where the surrounding earth below can no longer support its weight.

The life and characteristics of any single sinkhole will be determined by the underlying geology. In soft, sandy soils, sinkholes

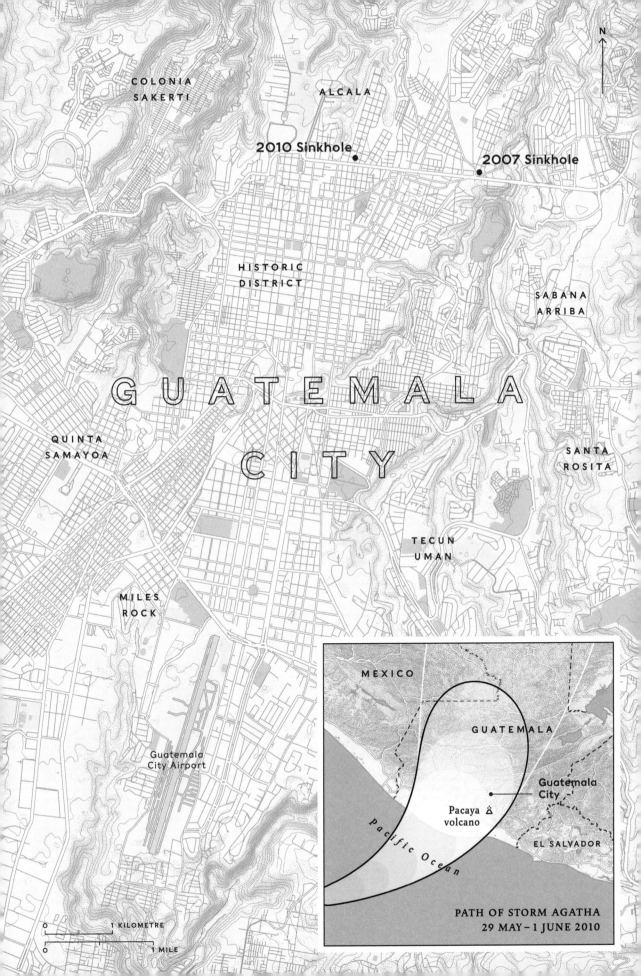

COLONIA
SAKERTI

ALCALA

2010 Sinkhole ●

2007 Sinkhole ●

HISTORIC
DISTRICT

SABANA
ARRIBA

G U A T E M A L A

C I T Y

QUINTA
SAMAYOA

SANTA
ROSITA

TECUN
UMAN

MILES
ROCK

Guatemala
City Airport

N

0 — 1 KILOMETRE

0 — 1 MILE

MEXICO

GUATEMALA

Guatemala
City

Pacaya Ӑ
volcano

Pacific Ocean

EL SALVADOR

PATH OF STORM AGATHA
29 MAY – 1 JUNE 2010

↑
Sinkholes have exposed
the fragility of the
foundations upon which
settlements such as
Guatemala City are built.

tend to be relatively shallow, as they don't have the strength to withstand much pressure before the surface material caves into the space created below. However, in firmer terrain such as karst limestone or clay, underground spaces slowly eroded by groundwater can grow considerably larger. When these eventually give in to the inevitable force of gravity — triggered by a range of potential destabilising factors, everything from heavy rain to extreme drought — they can expose large cavities. These can be capable of devouring people, cars and — as in Guatemala — even buildings.

News reports from the 2010 event focused on the immense size of the sinkhole that had opened in the city, roughly 20 metres (65 feet) across and up to 60 metres (200 feet) deep, alongside the almost perfectly round chasm that had been neatly holepunched through the capital's streets. With limestone not a factor in the unexpected appearance of this sinkhole, experts instead pointed to Guatemala City's history of building on top of loose volcanic pumice, and other debris deposited by centuries of eruptions. With this material essentially free to move around beneath the paved concrete, an incident such as this was almost inevitable. Indeed, it was the second such sinkhole to appear in the city in just three years — in February 2007, a comparably large hole emerged not far away, also during high rainfall.

It didn't take long for the experts drafted in to investigate these incidents to begin pointing fingers at poor-quality construction and an absence of building regulations. This sinkhole, they argued, was primarily caused by an infrastructural failure, most likely city sewers or storm drains rupturing due to the volume of water dumped by the tropical storm (not helped by the recent eruption of nearby volcano Pacaya, clogging up the drains with ash). This would technically make it a man-made event, and therefore not a proper sinkhole at all. This is likely to be little comfort to the residents of Guatemala City, with the prospect of suddenly disappearing into the earth an ever-present threat.

Svalbard Global Seed Vault

A VAULT TO PROTECT CROPS
FROM AROUND THE WORLD

NORWAY

N 78° 14′ 10″
E 15° 29′ 28″

War. Famine. Pestilence. The collapse of the global food supply could come in a variety of truly horrific and terrifying ways. If and when it does, feeding the world's population could be immensely difficult, given the globalised trading network necessary in modern agriculture, and, most importantly, the delicate supply of seeds required to grow the crops we need.

Edible crops don't simply appear by magic. A glance at the tough, seed-packed wild fruit that is the ancestor to the modern soft and sweet banana, for example, gives us an idea of how intentionally genetically modified our everyday contemporary crops are compared to their predecessors. Only careful breeding and domestication — sometimes over thousands of years, by indigenous communities all over the world — have given us the garden-variety fruits and vegetables we enjoy in the modern world, from almonds and sweetcorn to carrots and watermelons, and many more.

But local and global threats — including invasive pests and diseases that slip across borders, a loss of suitable growing environments, drought, extreme salinity and many more — endanger these developments. Ensuring the long-term survival of agricultural and industrial crops, and maintaining robust food security, are, many experts argue, of the utmost importance. With this in mind, many countries and regions have secure seed banks, places to store local seeds for the future in case the worst should happen.

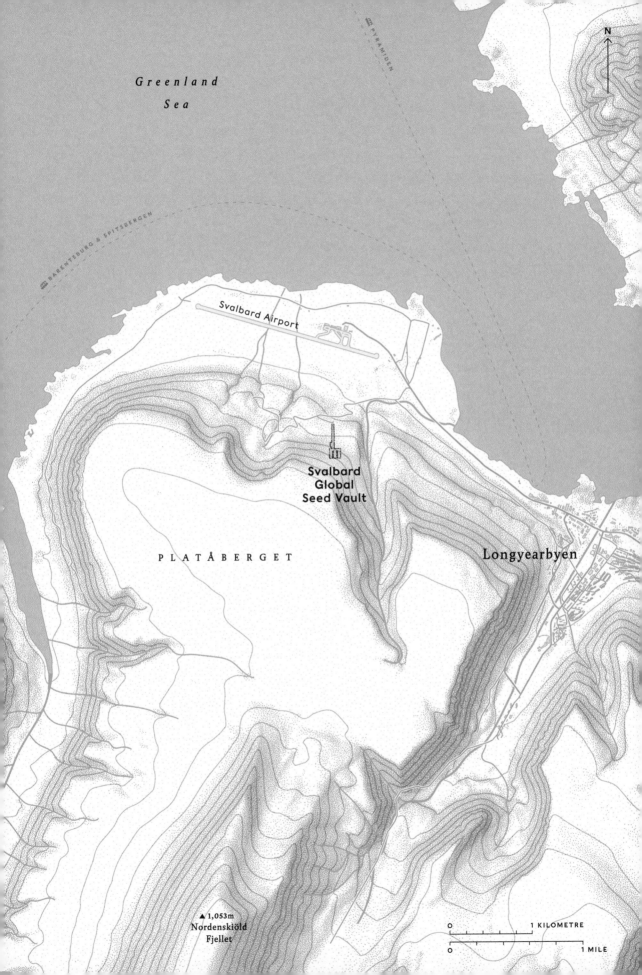

Greenland
Sea

N

⌖ PYRAMIDEN

⌖ BARENTSBURG & SPITSBERGEN

Svalbard Airport

Svalbard Global Seed Vault

PLATÅBERGET

Longyearbyen

▲ 1,053m
Nordenskiöld
Fjellet

0   1 KILOMETRE

0   1 MILE

But what if these banks themselves should become compromised, and their precious contents be lost? Then a back-up is required. In 1984, a 'failsafe' seed-storage facility was opened by the Nordic Gene Bank in an abandoned coal mine on the remote Norwegian Arctic archipelago of Svalbard. It was a noble effort in principle, but over time there was concern that a facility so potentially important to the survival of our species was being housed in a site that periodically experienced extremely high levels of hydrocarbon gases. A new location was required.

To great fanfare, the new Svalbard Global Seed Vault was opened in 2008 by the now-renamed Nordic Genetic Resource Centre (NordGen), operating in partnership with the Norwegian Ministry of Agriculture and the Crop Trust. This remote, heavily locked vault, constructed within a sandstone mountain 130 metres (426 feet) above the current sea level, acts as a free-of-charge last resort for the world's seed diversity, in case the network of regional and national seed banks around the world suffer catastrophic failures. A carpological time capsule, frozen deep within the permafrost and glaciated landscape of the Svalbard archipelago, 1,300 kilometres (800 miles) inside the Arctic Circle.

The vault is kept mechanically cooled to -18°C (just below zero degrees Fahrenheit) to ensure low metabolic activity. It is hoped that, even if the cooling system were to fail, the naturally glacial environment of Svalbard would maintain a sufficiently frigid temperature to prevent damage to the valuable resources inside.

For such an important facility, it appears remarkably spartan. The whole operation is designed to function with minimal need for the presence of humans, with staff appearing only on days when fresh deposits are scheduled to arrive. Inside the imposing outer doors, large industrial monochrome tunnels with curved concrete and metal walls slope downwards, through several more heavily sealed doors, towards the main vaults, buried over 100 metres (328 feet) inside the mountain.

The ambition of the project remains a wonder to behold. Over a million seed samples are contained within the 3,000 diligently

↑
Only the entrance to the Global Seed Vault is visible from outside, with the rest buried deep inside the bowels of the mountain.

Seeds from across the globe are held securely in subterranean Svalbard, where they are kept protected by the remote, icy environment.

categorised, carefully sealed boxes within the primary vault, comprising more than 6,000 different plant species. Impressive though these numbers are, the total capacity of the facility is far larger, with two as-yet-unused vaults giving the site the capacity to store up to 4.5 million crop varieties. At around 500 seeds per crop, this means there is space for an incredible 2.25 billion seeds within this single building.

In the words of the Crop Trust, the vault contains 'the most diverse collection of food crop seeds in the world'. Seeds from close to every country on Earth are represented down here, including a range of dietary staples such as wheat, maize, rice, barley, sorghum, potatoes, chickpeas, peanuts, oats and beans, in addition to the wild varieties of many of these crops. This diversity is reflected in the eclecticism of the depositors themselves — from wild carrots and orchids provided by the United Kingdom's Royal Botanical Gardens at Kew, to sacred corn and beans from the indigenous Cherokee Nation in the United States.

Interestingly, the vault does not (and will not) include any so-called 'genetically modified' crops, which are banned from import and/or storage by Norwegian domestic law. To date, there has been only one major incident where a depositor has request-ed to withdraw previously deposited seeds — in 2015, when the International Center for Agricultural Research in the Dry Areas (ICARDA) lost access to their bank in the Syrian city of Aleppo, and needed to restock their stores of wheat, lentils, chickpeas and other crops to begin populating their new facilities in Morocco and Lebanon.

As if the remote location and locked doors weren't considered sufficiently tight protection, there is also an unofficial security force in the form of wandering polar bears. But even that may not be enough. In late 2016, Arctic temperatures far above normal produced unexpected heavy rainfall on Svalbard (in lieu of the anticipated light snowfall), meaning the permafrost usually sur-rounding the Global Seed Vault was replaced by encroaching meltwater. This broke through the outer doors, and flooded the inner tunnel, where it froze to ice. On this occasion the seed vault itself, buried deep inside the mountain, survived intact. While a €20 million upgrade has since waterproofed the tunnel against possible repeat incidents, the vault will have to remain sturdy to withstand everything the future — especially the growing impact of climate change — will throw at it.

Helsinki
Underground City

AN UNDERGROUND URBAN LANDSCAPE
THAT DOUBLES UP AS A CITY-WIDE BUNKER

FINLAND

N 60° 10' 31"
E 24° 56' 07"

Sirens wailing, lights flashing, people running. In a future scenario where Finland and its capital, Helsinki, find themselves under attack and/or invasion, such a scene may well unfold. But the Finns wouldn't be blindly fleeing for the border. Instead, a carefully laid out underground shelter awaits, large enough to house the entire city population.

Finns are understandably wary of their neighbours. The four-month so-called 'Winter War', from late 1939 to early 1940, may not be widely known outside Russia and Finland, the two key combatants (Russia being the Soviet Union at the time) — muddled in, as it was, with the opening skirmishes of the Second World War — but it left an impression upon the Finnish consciousness that has arguably worked its way into the national psyche. Despite deploying over a million soldiers in the direction of the 1,309-kilometre (813-mile) Finnish border, the Soviets failed to make a significant break into their neighbour's territory, thanks to the dogged and skilful tactics of the unified and defensive Finns, helped by the poor decision to stage such a large military intervention during the harsh Nordic winter. Some Finnish territory was lost, but the majority of the country was saved, and a full invasion averted.

In the post-war years, while Finland may have joined their European neighbours in multilateral international organisations

N

KIVIHAKA

KUMPULA

RUSKEASUO

PASILA

LAAKSO

ALPPILA

HERMANNI

Olympic
Stadium

TÖÖLÖ

KALASATAMA

KALLIO

H E L S I N K I

Hietaniemi

Central
Station

Parliament
House

Helsinki Zoo

Senate Square

KAMPPI

Office of
the President

RUOHOLAHTI

PUNAVUORI

Helsinki
Observatory

KAIVOPUISTO

JÄTKÄSAARI

EIRA

MUNKKISAARI

LÄNSI-
MUSTA

Baltic Sea

Underground facilities, existing

Underground facilities, planned

0   1 KILOMETRE

0   1 MILE

such as the European Free Trade Association (EFTA) and the European Union (EU), it chose not to become a member of the more military-focused North Atlantic Treaty Organisation (NATO). Instead, the Finns opted for a tentative neutrality between the competing forces of East and West, even as former Soviet states such as the nearby Baltic nations of Estonia, Latvia and Lithuania willingly signed up to NATO. A fresh invasion from contemporary Russia would therefore prompt no automatic NATO response, and in theory would leave the Finns, once again, fending for themselves.

Hence the understandably alluring appeal of a huge bunker perched on the edge of the Baltic Sea, beneath the frigid streets of Finland's capital city. In the early 2000s, the Helsinki City Planning Committee and other administrative bodies began official consultations regarding a potential expansion of the existing, but relatively small, underground spaces that had been maintained below Helsinki since the 1980s. The consequential strategy spent nearly a decade in planning, but by the end of 2010, Helsinki City Council was able to unveil what had become known as the Underground Master Plan, the first such city-wide subterranean plan in the world. Including 200 kilometres (124 miles) of passageways, shelters and infrastructure (such as vital water and internet capabilities) dug into the bedrock and spreading out from below the city centre like a plate of spaghetti, it enables the whole city — over 600,000 residents — to quickly become subterranean, were an invasion to ever occur.

All this is very exciting and headline-grabbing, but Finland's underground city is more than a glorified hiding place. During peacetime — to date, all of the time — the purpose of these underground amenities is to be part of a sophisticated urban plan to move public and private facilities, such as shops and churches, underground (in part to overcome the long Baltic winters, when Helsinki can receive as little as six hours of sunlight a day). As well as the public transportation, sewage systems and power cables found beneath many capital cities, Helsinki's buried facilities now include sports pitches and car parks, large open spaces for which

Subterranean facilities such as churches and leisure centres are part of Helsinki's transition to a hybrid, part-underground capital.
↓

it would be difficult to find room above ground. It is estimated that Helsinki now has 1 square metre of underground capacity for every 100 square metres above ground, and growing. And, of course, many of these can be quickly repurposed as emergency shelters should the unthinkable ever happen.

Kolwezi Mines

WHERE MINERS RISK THEIR LIVES SOURCING ESSENTIAL METALS FOR MODERN GADGETS

DEMOCRATIC REPUBLIC OF THE CONGO

S 10° 42' 22"
E 25° 26' 32"

The average smartphone contains around 31 grams of aluminium, 19 grams of iron and 8 grams of copper, alongside traces of many other materials, such as gold, tungsten and gallium. Building such technologically sophisticated machines requires the careful integration of specific resources brought together from around the world. A similar story can be told about other highly advanced devices symbolic of the twenty-first century, everything from the tablet computer to the electric car. Of these essential materials, perhaps none is more vital, contested and dangerously unregulated than cobalt.

Cobalt is a crucial component of lithium-ion batteries, with around 7 grams in your average smartphone. But the sourcing of this material is contentious to say the least. From the end point of the glamorous high-street retailers who flaunt these sleek devices, via the Chinese factory cities of Shenzhen, Guangzhou and others where they are manufactured, we need to travel back thousands of miles, across the Indian Ocean, via the seaports of Tanzania and South Africa, to the origin of the material that powers them. Here, in the southeast corner of the Democratic Republic of the Congo (DRC), lies the dirty secret behind these shiny gadgets.

The reality of the cobalt mines in the DRC is very different from the utopian ideals that are commercially associated with the

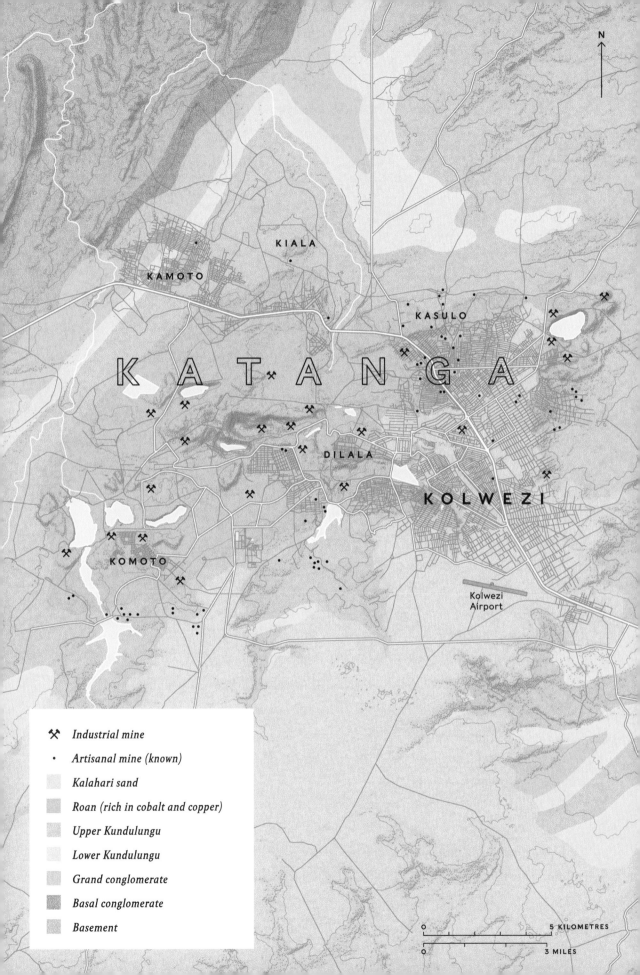

N

KIALA

KAMOTO

KASULO

KATANGA

DILALA

KOLWEZI

KOMOTO

Kolwezi
Airport

⚒ Industrial mine

• Artisanal mine (known)

Kalahari sand

Roan (rich in cobalt and copper)

Upper Kundulungu

Lower Kundulungu

Grand conglomerate

Basal conglomerate

Basement

0   5 KILOMETRES

0   3 MILES

devices that feed off them. Viewed from above, it appears that the land has been hacked apart, huge gashes streaking across the landscape. Here is where a significant proportion of the world's cobalt is obtained — in the form of the oxidised ore heterogenite — before being sold into the global marketplace, and eventually into the hands of some of the world's richest companies.

With around half the global reserves, the DRC produces 60 per cent of the world's annual supply of cobalt, much of which is concentrated in Kolwezi and other regions in the south. The industry is dependent on the metals scraped from the earth in this remote part of central Africa by an estimated 150,000 (although likely many more) informal workers known as 'creuseurs'. DRC cobalt experts believe around a fifth of the country's exports is obtained from the labour of these workers and their rudimentary mining efforts. And as demand has risen, so has the price of cobalt, enticing more and more people in the DRC to try to make money from scratching around in the dirt.

A trip down one of the artisanal mines in Kolwezi reveals this hard and unforgiving world. Muddy tunnels, crudely dug deep into the red dirt. They are dark, full of toxic dust, and, for those who work there, the perennial fear of flooding, fatal collapse or being buried alive. A thin pipe attached to a small, dirty ventilator is entrusted to provide fresh air to overpower the methane and other dangerous gases that accumulate in these tunnels. Down here, tens of thousands of creuseurs — poverty-stricken men and boys — work twelve- to twenty-hour days, for as little as a dollar or two a day. They are without safety protection or shoes, without maps or proper tools, without adequate training or company support, and without access to medical care once they experience sickness or poisoning — caused by the very materials they are down here searching for.

Those without access to tunnels can try their luck digging (often with bare hands) through the discarded by-products of the industrial mines, an equally hazardous line of work. Women and children then spend hours washing the raw mined materials in filthy, toxic rivers, back-breaking labour for which they too receive

a pittance (and which has been linked to a rise in birth defects). Campaigners describe the work as modern-day slavery.

There are many places in the world where subterranea holds uncomfortable truths about which the rest of the world might prefer not to know. Kolwezi is not alone. But certainly the images and stories that have emerged from the DRC paint one of the most striking pictures of how dependent contemporary, glossy twenty-first-century life is on raw materials dug from deep below earth, sometimes in harrowing conditions. Some things that take place underground should be exposed to the light, whether or not we like what is revealed.

Jerusalem Cemeteries

A NEW UNDERGROUND CEMETERY
TO EASE CONGESTION

ISRAEL

N 31° 47′ 51″
E 35° 10′ 27″

The underground world beneath the city of Jerusalem could — and indeed does — fill many literary works. When one is a sacred site for three major Abrahamic religions (Islam, Christianity and Judaism) covering more than half the global population, one tends to have accumulated more than one's fair share of historical architecture and artefacts. Buried beneath the modern high-rise office blocks, shopping malls, technology parks and gated communities that make up this congested, contested, auspicious, delicious Middle Eastern metropolis lie many fascinating subterranean spaces. The pioneering (if crude) excavations ordered by British archaeologist Charles Warren in the late 1860s may be among the most well known, especially around the renowned Temple Mount. But further discoveries have continued to surface during the century and a half since then — from Jewish baths to Roman quarries to a 600-metre (2,000-foot) street lost for nearly 2,000 years — spread out beneath the old city's four traditional quarters.

Globally, subterranea is of course a common destination for the dead. But in Jerusalem, this traditional route has become a real traffic jam. Historically, the hallowed Mount of Olives, a sanctified site that is an important place for Jewish pilgrims (plus a popular lookout spot for regular visitors to the city), has been the top burial site of choice for many. Yet, like every cemetery, this

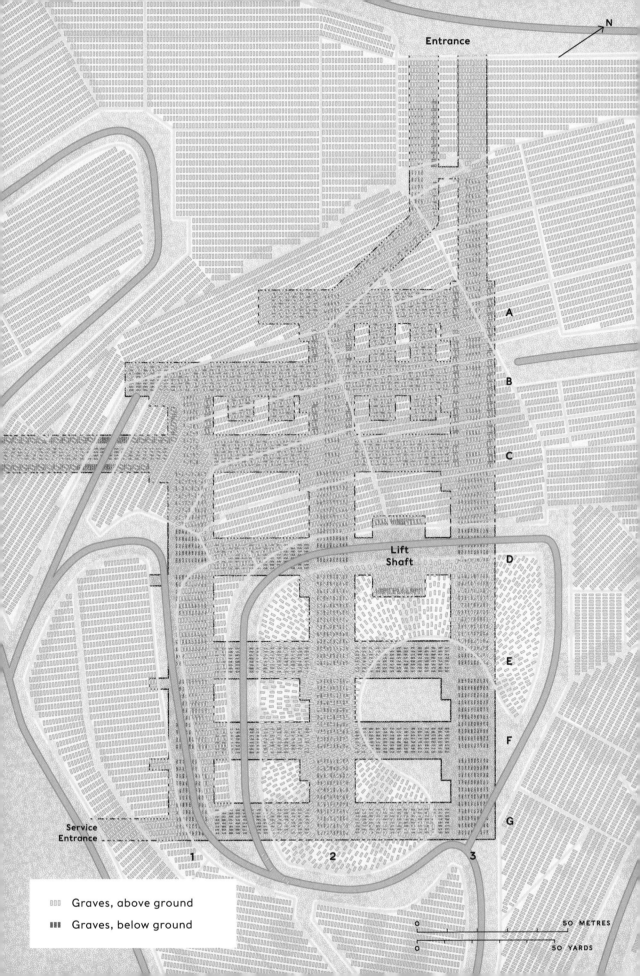

Entrance

N

A

B

C

Lift
Shaft

D

E

F

G

Service
Entrance

1   2   3

Graves, above ground

Graves, below ground

0   50 METRES

0   50 YARDS

one has a spatial limit, and with over 120,000 recorded graves, it is close to reaching maximum occupancy. Expansion is unlikely, especially as the Mount of Olives is situated in the ever-disputed East Jerusalem, making it vulnerable to vandalism and geopolitical tensions. Most other cemeteries still accepting bodies in Jerusalem are either impractically small or reserved exclusively for notable individuals such as political or military leaders, for example Mount Herzl. With space-saving cremation not a realistic option for the city's majority Jewish population, and the few remaining burial plots selling for up to US $20,000 a pop, the need for more space has become critical.

'Faith hovers over the towers of Jerusalem,' commented

↑
The iconic Mount of Olives draws pilgrims and visitors from around the world, but is now at near capacity.

nineteenth-century British Prime Minster Benjamin Disraeli. And there are certainly now many more such towers, since the initial solution to the burial crisis was to go up. In the previous century, burial towers were erected en masse on the outskirts of the city. Har Hamenuchot, the largest such installation, looms over the highway that connects Jerusalem with its Mediterranean sibling, Tel Aviv. Opened in 1951, it now holds more than 150,000 deceased individuals, and is growing fast. The 'success' of Har Hamenuchot and others saw the Jerusalem Jewish Community Burial Society expand the scheme, with dominating towers full of resting souls rising from the desert to scrape the skies outside settlements across Israel.

But these vertical cemeteries were not popular with everyone, especially given the prominent place many of them take in the skylines of Jerusalem. Therefore, authorities have now taken an abrupt U-turn, instead heading down, deep into the guts of the city. In partnership with Israeli tunnelling firm Rolzur, a new facility has recently been constructed in the clay soils beneath Har Hamenuchot: a modern, subterranean, three-floored, twelve-tunnelled tomb with space for 23,000 people, and possibly many more in the future. Unlike the chaotic medley of graves that blanket the surrounding hills up at surface level, this underground facility is carefully ordered, with neat horizontal and vertical rows dug out in a grid fashion. The decor is suitably futuristic, with sleek passageways, large elevators, regular ventilation and soft lighting. The departed can now spend the afterlife in comfort and style.

To date, the experiment appears popular, with funding for the initial facility obtained courtesy of prepayment from people keen to secure their future beyond the mortal realm in this experimental new space. And while a reported US $70 million price tag might appear steep, developers will hope that the millions of Jewish followers both in and outside Israel will continue to be interested in investing in a spot in the revered turf of the homeland upon their eventual passing.

Jerusalem appears to be at a pivot point, where the city ceases to witness towers scaling ever further towards the skies, and instead copies Har Hamenuchot's example, tunnelling down, adding yet further to the catacombs of history contained beneath the city streets.

Could this state-of-the-art, freshly constructed underground cemetery solve Jerusalem's burial problems?

→

Concordia
Research Station

COLLECTING ICE CORES FROM GLACIERS
AROUND THE WORLD BEFORE THEY MELT AWAY

ANTARCTICA

S 75° 06' 02"
E 123° 20' 05"

Storing ice underground in Antarctica, in a cavern built from ice, might sound like a pointless activity, given that the South Pole frankly has plenty of ice going spare, thank you very much. But not all ice is the same. In fact, from a scientific perspective, some ice is close to priceless. In a best-case scenario, it can provide deeply valuable empirical data about our past, present and, quite possibly, even future.

The ice core is a staple of glacial and climate modelling, particularly in the specialised field of palaeoclimatology. Over time, as layers of snow and ice build up on ice sheets and glaciers, they trap particles such as dust, ash and pollen, plus small bubbles of air. In expert hands, these tiny, sometimes microscopic, specimens can reveal immensely useful information about the gaseous composition of past atmospheres and environments at the time when that layer of ice was being formed. This is done by measuring carbon dioxide and methane levels, the characteristics of frozen bacteria, and the isotopes of the trapped water itself.

These cores can stretch back tens of thousands, or even hundreds of thousands, of years. The oldest core ever obtained, from Antarctica, includes ice that is 800,000 years old. In the words of NASA, 'ice cores have proven to be one of the most valuable climate records to date'. And when you know what occurred in the past, it becomes considerably more feasible to predict what

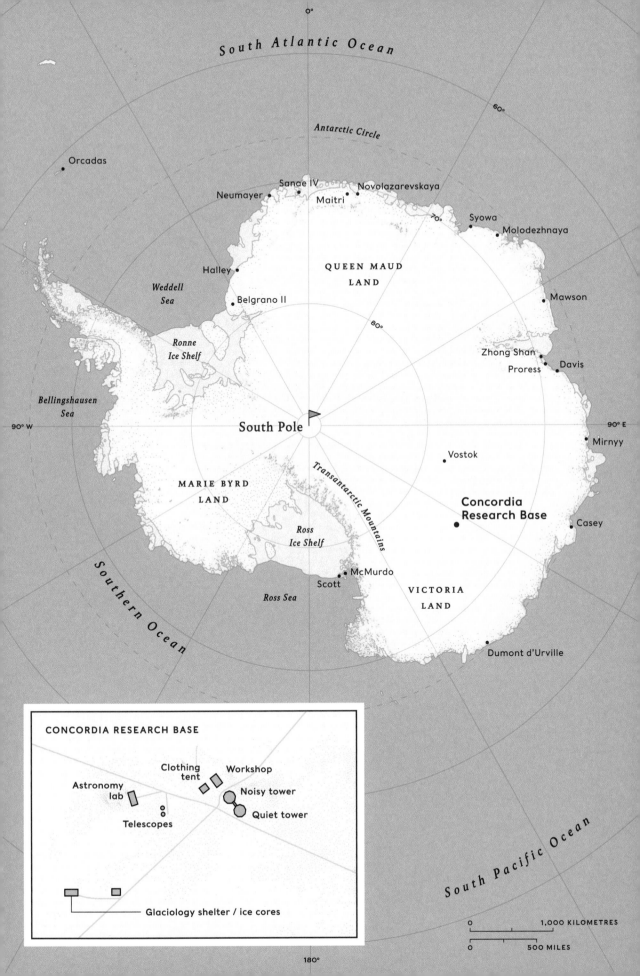

South Atlantic Ocean

0°

Antarctic Circle

60°

Orcadas

Neumayer   Sanae IV   Novolazarevskaya
Maitri
Syowa
70°
Molodezhnaya

QUEEN MAUD
LAND
Halley
Weddell
Sea
Belgrano II
Mawson

Ronne
Ice Shelf
80°
Zhong Shan
Proress
Davis

Bellingshausen
Sea

90° W
South Pole
90° E
Mirnyy

Vostok

MARIE BYRD
LAND
Concordia
Research Base
Casey

Ross
Ice Shelf

Transantarctic Mountains

Southern
Ocean
McMurdo
Scott
VICTORIA
LAND
Ross Sea

Dumont d'Urville

CONCORDIA RESEARCH BASE

Astronomy
lab
Clothing
tent
Workshop
Noisy tower
Quiet tower
Telescopes

Glaciology shelter / ice cores

South Pacific Ocean

0   1,000 KILOMETRES
0   500 MILES

180°

↑

Keeping ice core samples in deep freeze is critical for allowing future palaeoclimatologists to study past climates.

might happen in the future, an especially helpful asset when facing an increasingly unstable climate.

And yet here climate change plays the role of both subject and saboteur, of defendant on trial and hired hitman. The very global problem these ice cores help scientists identify is the same one that threatens to wipe humanity from existence, like a house fire symbolically destroying a fire alarm. Half of the ice within the 4,000 glaciers in the Alps is forecast to melt away by the year 2050, and up to two-thirds by the end of the century. In the Himalayas, up to a third of all glacial ice is believed to already be irreversibly lost, with another third forecast to disappear by 2100.

While these time frames may seem distant, and while the valuable lower layers of ice may not all be melting yet, liquid water on the surface is capable of percolating down through a glacier and contaminating lower levels, mixing up samples to the point where they become scientifically unreliable, even useless. It's therefore a race against time to deduce as much information as possible from the historical records of these glaciers before they are lost for ever.

Enter the Ice Memory Project: an underground vault made of ice, dug inside a snow cave beneath Antarctica's Franco-Italian Concordia Research Station (one of only three permanent Antarctic bases situated away from the continent's coastline). Considered the world's first library of archived glacier ice, the project was launched in August 2016 with the transportation of three cores 130 metres (427 feet) long — the height of Egypt's Great Pyramid of Giza — from the Col du Dôme glacier on Mont Blanc, France, via helicopter, to a cold-storage facility in nearby Grenoble.

Here, one core was kept for analysis, while the other two were loaded onto a ship to take them to their new subglacial home only a thousand miles from the South Pole, where average temperatures at a frostbiting -54°C (-65°F) will protect them for analysis many years into the future. Similar cores are now being collected from other melting glaciers around the world, such as in Russia and Bolivia, with countries such as Germany, Austria,

Switzerland, the United States, China, Nepal and Canada all expressing an interest in utilising the secure storage facility at Concordia.

The underground conditions in which this icy vault operates, where the cores are kept, are, like most things in Antarctica, unglamorous. A temporary cave is carved inside the snow that, once hardened, will serve as a reliably cold-storage fridge for the next decade or so, to be re-carved once the icy walls eventually collapse. The entrance is small and foreboding, the interior dark, the environment predictably frigid, and the storage containers themselves mere industrial boxes. But the people working here, trudging through the snow with packages of ice core samples, are driven by a powerful motivation: the hope that one day, new, as-yet-unrealised technologies will be capable of extracting further valuable data from these precious cores about past climates. With a new, warmer, more volatile climate looming, they may be vital in anticipating, preparing for and mitigating the worst of what the future has in store.

The Aurora Australis
flickers over the
Concordia base, situated
only a thousand miles
← *from the South Pole.*

Los Angeles Tunnels

A FUTURISTIC TRANSPORT PROPOSAL
THAT IS STEEPED IN HISTORY

UNITED STATES

N 34° 00′ 07″
W 118° 19′ 58″

In the cult 1980s sci-fi movie *Blade Runner*, a futuristic Los Angeles is depicted as a dystopian high-rise city where elite residents travel in flying cars high above the ground. And perhaps understandably so, because in the real world, gridlock grips LA on a daily basis. In 2019, the year in which the movie is set, the suntanned residents of Los Angeles are calculated to have each lost 103 working hours being stuck in traffic, burping toxic fumes into the hot Californian air. It's no surprise that helicopters are often seen flying over the heads of the road-raged on the boulevards below — just like the wealthy and powerful characters in *Blade Runner* — enabling those with especially deep pockets to quickly go about their business without being stuck on the highways with the rest of the frustrated Angelenos.

It was eventually all too much for one billionaire. 'Traffic is driving me nuts,' tweeted Elon Musk, founder of electric car manufacturer Tesla and rocket company SpaceX, in December 2016. 'Am going to build a tunnel boring machine and just start digging…' What was perhaps initially a social media announcement posted with his tongue firmly lodged inside his cheek, eventually, somewhat inexplicably, became a functioning company. The following year, Musk announced that what he had named 'The Boring Company' would soon set to work, undertaking his stated goal of burrowing beneath the choked streets of a number of major US cities, including Las Vegas, Chicago and Baltimore,

N

SUN VALLEY

Verdugo Mountains

Sherman
Oaks

BURBANK

NORTH
HOLLYWOOD

PASADENA

Hollywood Hills

Vermont /
Sunset

The Getty

BEVERLY
HILLS

HOLLYWOOD

Dodger
Stadium

UCLA / Westwood

Echo Park /
Silver Lake

Union
Station

L   O   S

West LA

USC /
Coliseum

Staples
Centre

King Eddy Saloon

EAST
LOS ANGELES

Santa
Monica

Leimert
Park

Venice /
Marina

A   N   G   E   L   E   S

Culver City

Inglewood

SOUTH GATE

LAX

LA Stadium

South LA

Hawthorne

COMPTON

Lawndale

South Bay

Torrance

TORRANCE

Carson

Long Beach
Airport

ROLLING
HILLS

— Tunnels, proposed

— Tunnels, in progress

0   5 KILOMETRES

0   3 MILES

to enable cars to be fired at speeds of up to 250 kph (155 mph) —
roughly the same speed as a skydiver plunging towards Earth — as
a way to circumvent above-ground traffic. Testing and early demon-
strations have shown glossy futuristic tunnels, with colourful strip
lighting and sleek Teslas cruising along inside them.

But it's in the City of Angels that Musk appears to feel the most
impact could be made, with prospective plans announced for a wide-
ranging network sprawling out beneath the city by the middle of
the century. Strangely, it's a move that harks back to an important
part of the city's heritage, to a less-automobiled time. In December 1925, LA's first subway lines were opened to great fanfare. They
were the final component of the city-wide Pacific Electric Railway,
which was attempting to replicate the subterranean public transportation train networks operating in New York, Boston and around
the world. It was only the mass expansion of highways over the subsequent decades that saw the gradual demise of this network and
enabled the car to become king (over 80 per cent of LA journeys
are now taken by private vehicle). Those empty concrete passageways, which once shuttled so-called Red Cars packed with thousands of commuters everywhere from Hollywood to Glendale, now
sit crumbling, dirty, coated in graffiti and thoroughly unloved.

Tunnels also played a key role in Los Angeles during the Pro-
hibition years of 1920 to 1933. Thanks to 18 kilometres (11 miles)
of secretive, under-utilised service passageways, smugglers and
beer barons were able to transport their alcoholic loot around
beneath the heavily policed streets, supplying illegal speakeasies with the valuable spirits that kept their clientele coming back
night after night. These venues have become legendary. The King
Eddy Saloon, a basement bar beneath the King Edward Hotel on
central Skid Row, whose street front claimed to be nothing more
than a humble piano store, is still connected to the tunnel that
used to supply liquor when imbibing such beverages was a felony.
Other hotels and seemingly prestigious establishments got in on
the act, with the tunnels even running underneath — and likely
supplying — courthouses and other government buildings. It's
evidence that backs up the rumours that bootlegging operations

↑
*Popularity for public
transport in Los Angeles
plummeted with the rise
of the private car in the
early twentieth century.*

during the dry era ran (metaphorically, at least) right through the mayor's office.

Of course, speakeasies and other supposedly unsavoury establishments never existed on any official register, and as such left no paper trail of when they opened or closed. This provides motivation for LA's urban spelunkers, who regularly spend their free time scouring around dingy underground spaces, in the hope they might stumble upon abandoned watering holes, lost for decades somewhere beneath glitzy Sunset Boulevard.

If Musk gets his way, and the city's underground tunnels become as well drilled and glossy as the teeth in a Hollywood smile, swapping decay for shiny new fillings, then the hustle and bustle of twentieth-century highways might well be shuffled to the lower levels. Perhaps then these decrepit underground spaces of bare brick walls and exposed concrete just might come back to life. If so, unlike the world envisaged in *Blade Runner*, where technology enables commuters to take to the skies, LA's transportational future might yet be rooted in subterranea.

Futuristic new tunnels will ease traffic congestion in Los Angeles, according to their billionaire creator.
↓

Hellisheidi

TECHNOLOGY THAT PROPOSES TO
BURY THE CLIMATE CHANGE CRISIS

ICELAND

N 64° 02' 39"
W 21° 23' 31"

Huge waves rise and fall in front of an intimidating wall of rugged, wild rock that emerges aggressively from the ocean, like ribs erupting from a corpse. This is the Reykjanes peninsula, Iceland. It's one of the few places along the Mid-Atlantic ridge — the continental divide that traverses the Atlantic Ocean, from the Arctic Circle to the Southern Ocean — that rises above the ocean waves. Combined with the explosive power of Iceland's volcanoes, it's no wonder that this raw terrain is where author Jules Verne decided to send his protagonists underground in the 1864 novel *Journey to the Centre of the Earth*. It's also not far from where a curious subterranean experiment is taking place that might yet save humanity from itself.

Prior to the Industrial Revolution of the nineteenth century, the carbon dioxide (CO_2) concentration of Earth's atmosphere throughout human history peaked at around 280 parts per million (ppm). But a century and a half of rapidly accelerating burning of coal, oil, gas and other energy-dense organic fuels has seen this concentration climb, first steadily, then close to exponentially. By 2020, this figure was recorded at up to 417 ppm, and climbing by a part or two annually. Experts believe the atmospheric concentration of carbon dioxide needs to be returned to below 350 ppm to prevent long-term climatic instability caused by the extra heat this gas traps in the atmosphere. Unfortunately, carbon dioxide

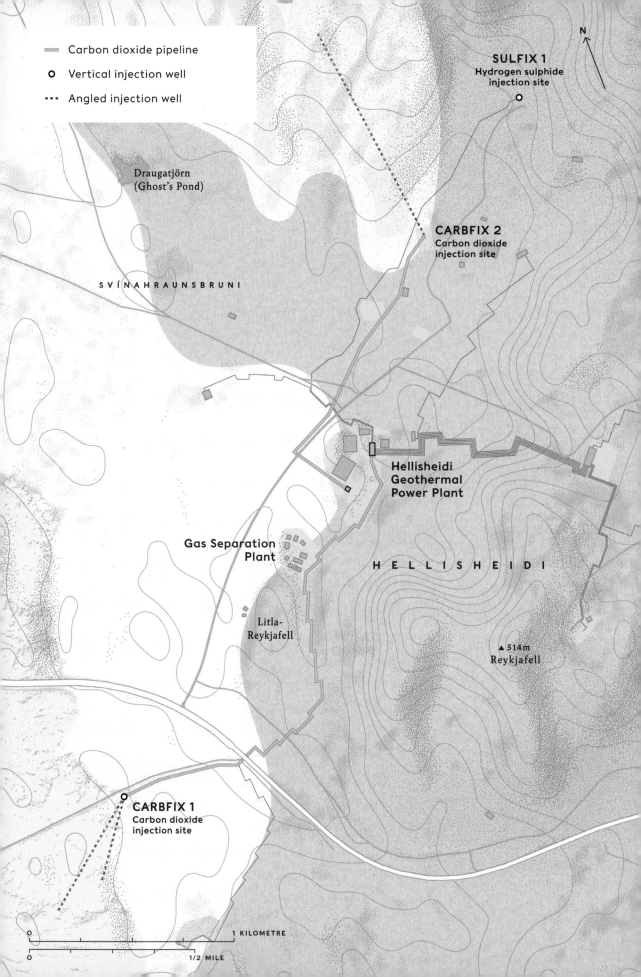

Carbon dioxide pipeline
○ Vertical injection well
··· Angled injection well

N

SULFIX 1
Hydrogen sulphide
injection site
○

Draugatjörn
(Ghost's Pond)

CARBFIX 2
Carbon dioxide
injection site

S V Í N A H R A U N S B R U N I

Hellisheidi
Geothermal
Power Plant

**Gas Separation
Plant**

H E L L I S H E I D I

Litla-
Reykjafell

▲ 514m
Reykjafell

CARBFIX 1
Carbon dioxide
injection site

○ |_____| 1 KILOMETRE
○ |_____| 1/2 MILE

can take centuries to naturally break down, so the accumulation has become a leading cause of the modern climate change crisis. Iceland is at the centre of one plan to reverse this.

Icelanders are already leading the way towards a low-carbon future through their mainstream use of geothermal energy, piped up from below ground, which releases much less carbon dioxide than other methods of power generation. But at Hellisheidi, Iceland's largest geothermal power plant, 25 kilometres (15 miles) east of the capital, Reykjavik, an experimental project aspires to go even further. Only a small fraction of the emissions from the plant are carbon dioxide — less than half a per cent — but when that small quantity emerges from below the earth, it is captured by an organisation named Carbfix. Instead of being allowed to freely enter the atmosphere, the gas is pumped into water, in a method similar to how fizzy drinks are made, then taken a few kilometres from Hellisheidi and pumped deep below ground, to around 800 metres (2,600 feet) below the surface. Here, in just two years — incredibly fast compared to the centuries or even millennia that natural processes normally take — the majority of the CO_2 becomes mineralised into stone; specifically, basalt. The carbon has been returned to the earth, instead of being allowed to wreak havoc in the atmosphere.

Carbfix began operations in 2012 and became a core component of the plant two years later. By the end of 2018, over 40,000 tonnes of carbon dioxide were being captured and buried by this process every year. But this was far from the world's first attempted project to find a way to send excess carbon down into the ground. The process known as carbon capture and storage (CCS) has been floated since the 1970s, proposed as a clever trick to allow for the continued burning of oil, gas and coal. But early attempts lacked the economic motivation and legislative robustness to make them into mainstream components of the energy industry. Despite the Intergovernmental Panel on Climate Change (IPCC) advocating CCS as a pivotal tool for mitigating the worst of climate change, by the early 2010s the technology appeared permanently stuck in second gear.

The key difference at Hellisheidi has been developing the ability to store carbon not as a gas, but as solid rock, making it much harder to accidentally leak. If successful at scale, such a process could potentially be utilised in power stations around the world, anywhere with access to the same basalt bedrock as Hellisheidi (as well as the regular water supply to enable pumping injections to take place). Washington and Oregon are two north-west US states that have begun experimenting with this model, while the idea is also being explored in central India. Through such an act of anthroturbation, humanity might be able to sweep the evidence of our extravagant behaviour since the Industrial Revolution under the metaphorical carpet. If it proves viable, it could enable the consequences of our vices to disappear deep underground.

Out of sight, out of mind.

Could the experiments taking place at Hellisheidi enable carbon dioxide to be hidden deep underground?
→

↑

Deep in the labyrinth beneath Waitomo, an eerie glow betrays the presence of hungry glow-worms, dangling sticky threads to capture unsuspecting prey (see page 3).

Acknowledgements

There is something undeniably mysterious, perhaps even slightly macabre, about the concept of the underground world that really stirs the imagination. I couldn't help noticing how easily people were engaged by the idea of *Subterranea* as soon as I mentioned it in casual conversation, many of whom then felt compelled to espouse at length about some fascinating subterranean place they had heard about, several of which then ended up in print. I'm therefore hugely appreciative of the many friends, family and colleagues — even passing acquaintances — who willingly suggested stories for the book: it would have been considerably more lightweight without your contributions.

Working with the very talented Matthew Young was nothing but a pleasure. His designs have made this book utterly gorgeous, from an eye-catching cover to some beautiful cartography. Even with the significantly more abstract chapters, he rose to the challenge and produced consistently excellent maps that I'm delighted to have my writing associated with. Shoaib Rokadiya and Lindsay Davies have been fantastic editors to work with, showing great patience and creativity as we collaborated to produce the highest-quality book we possibly could. Sincere thanks go to Cathie Arrington for her hard work researching and licensing all our images, as well as to Alex Clarke and the rest of the team at Wildfire. There is also a long list of academics, explorers, experts and others who contributed time and wisdom to ensure the accuracy of the book, and many photographers whose talent helped make it shine visually. I'm immensely grateful to you all.

As a lifelong claustrophobe, this book was less an impassioned personal tribute to small dark spaces and more a celebration of the astonishing diversity of subterranean places to be found in and around this endlessly fascinating planet we call home. With that in mind, I'm incredibly grateful to Mum and Dad for nurturing in me a relentless geographical curiosity that has proved very useful in this and other projects. No matter where my personal or professional life has taken me, they have never been anything except supportive and encouraging. Many thanks also go to my sister, Charlotte, of whom I am immensely proud, and to my wonderful Grandma, a constant source of joy and inspiration. Finally, *muito obrigado* to my partner, Ana, for her love, endless positivity, and for accompanying me every step of the journey. *Te amo.*

Further Reading

The stories in *Subterranea* consist of facts and anecdotes from hundreds of books, articles, videos, radio broadcasts, websites and databases, and I'm especially grateful for the existence of *Encyclopedia Britannica* and countless hard-working news organisations. While most chapters amalgamate the most interesting details from this wide range of sources, I simply couldn't have told certain stories without some key references, and therefore I must credit the following:

— Allred, Kevin; Allred, Carlene: 'Development and Morphology of Kazumura Cave, Hawaii', *Journal of Cave and Karst Studies*, August 1997

— Amnesty International/Afrewatch: 'Democratic Republic of Congo: "This is what we die for": Human rights abuses in the Democratic Republic of the Congo power the global trade in cobalt', 19 January 2016, https://www.amnesty.org/en/documents/afr62/3183/2016/en/

— Bisharat, Andrew: 'Epic flood sends cavers scrambling for their lives', *National Geographic*, 18 October 2018, https://www.nationalgeographic.com/adventure/2018/10/flood-escape-deepest-cave-veryovkina-abkhazia/

— Brooks, Darío: 'La Cueva de los Tayos, la legendaria y misteriosa formación de Ecuador que despertó la fascinación del astronauta Neil Armstrong', *BBC Mundo*, 27 November 2017, https://www.bbc.com/mundo/noticias-42104844

— Camille Aguirre, Jessica: 'The Story of the Most Successful Tunnel Escape in the History of the Berlin Wall', *Smithsonian Magazine*, 7 November 2014, https://www.smithsonianmag.com/history/most-successful-tunnel-escape-history-berlin-wall-180953268/

— Doel, Ronald E.; Harper, Kristine C.; Heymann, Matthias: *Exploring Greenland: Cold War Science and Technology on Ice*, Palgrave Macmillan US/Springer Nature, New York, 2016

— Fox-Skelly, Jasmin: 'Once a year, people poison these fish as part of a ritual', *BBC Earth*, 14 April 2016, http://www.bbc.com/earth/story/20160413-the-fish-that-swims-in-toxins-and-gets-poisoned-by-humans

— Frankel, Miriam: 'Religious rite gives evolution a helping hand', *New Scientist*, 14 September 2010, https://www.newscientist.com/article/dn19447-religious-rite-gives-evolution-a-helping-hand/

— Fredrick, James: '500 Years Later, The Spanish Conquest of Mexico Is Still Being Debated', *Weekend Edition Sunday*, NPR, 10 November 2019, https://www.npr.org/2019/11/10/777220132/500-years-later-the-spanish-conquest-of-mexico-is-still-being-debated

— Grove, Thomas: 'Beneath Helsinki, Finns Prepare for Russian Threat', *Wall Street Journal*, 14 July 2017, https://www.wsj.com/articles/beneath-helsinki-finns-prepare-for-russian-threat-1500024602

— Hanbury-Tenison, Robin: *Finding Eden: A Journey into the Heart of Borneo*, I. B. Tauris, Bloomsbury Publishing Plc., London, 2017

— Hansen, James: *First Man: The Life of Neil A. Armstrong*, Simon & Schuster, New York, 2018

— Hennessy, Peter: *The Secret State: Preparing for the Worst 1945–2010*, Penguin, London, 2010

— Jay Deiss, Joseph: *Herculaneum: Italy's Buried Treasure*, Getty Publications, Los Angeles, 1989

– Jenner, Andrew: 'Get Lost in Mega-Tunnels Dug by South American Megafauna', *Discover Magazine*, 28 March 2017, https://www.discover magazine.com/planet-earth/get-lost-in-mega-tunnels-dug-by-south-american-megafauna

– Last, Alex: 'Vietnam War: The Cu Chi Tunnels', *Witness History*, BBC World Service, 3 January 2017, https://www.bbc.co.uk/programmes/p04kxnbt

– Mace, Fred: 'Account of Discovery of Waitomo Caves', *King Country Chronicle/Waitomo News*, Te Kuiti, 1 October 1910, https://paperspast.natlib.govt.nz/newspapers/KCC19101001.2.4.2

– Mangold, Tom; Penycate, John: *The Tunnels of Cu Chi: A Remarkable Story of War*, Weidenfeld & Nicolson, London, 2012

– Marzeion, Ben; Levermann, Anders: 'Loss of cultural world heritage and currently inhabited places to sea-level rise', *Environmental Research Letters*, 4 March 2014, DOI: 10.1088/1748-9326/9/3/034001

– Matter, Juerg M et al: 'Rapid carbon mineralization for permanent disposal of anthropogenic carbon dioxide emissions', *Science*, 10 June 2016, DOI: 10.1126/science.aad8132

– Neumann, Joachim: 'Experience: I tunnelled under the Berlin Wall', *Guardian*, 12 July 2019, https://www.theguardian.com/world/2019/jul/12/experience-i-tunnelled-under-the-berlin-wall

– Nunez, Christina: 'Q&A: The First-Ever Expedition to Turkmenistan's "Door to Hell"', *National Geographic*, 17 July 2014, https://www.nationalgeographic.com/news/energy/2014/07/140716-door-to-hell-darvaza-crater-george-kourounis-expedition/

– Otman, Waniss; Karlberg, Erling: *The Libyan Economy: Economic Diversification and International Repositioning*, Springer-Verlag Berlin Heidelberg, Berlin, 2007

– Palmer, Jane: 'Why ancient myths about volcanoes are often true', *BBC Earth*, 18 March 2015, http://www.bbc.com/earth/story/20150318-why-volcano-myths-are-true

– Rogers, Paul; McAvoy, Darren: 'Mule deer impede Pando's recovery: Implications for aspen resilience from a single-genotype forest', *PLoS ONE*, 17 October 2018, DOI: 10.1371/journal.pone.0203619

– Ruggeri, Amanda: 'The strange, gruesome truth about plague pits and the Tube', *BBC Autos*, 6 September 2016, http://www.bbc.com/autos/story/20160906-plague-pits-the-london-underground-and-crossrail

– Simon, Matt: 'Fantastically Wrong: The Legendary Scientist Who Swore Our Planet Is Hollow', *WIRED*, 2 July 2014, https://www.wired.com/2014/07/fantastically-wrong-hollow-earth/

– Songwriter, Jason: 'The sweet spot for building drug tunnels? It's in San Diego's Otay Mesa neighborhood', *Los Angeles Times*, 22 April 2016, https://www.latimes.com/local/lanow/la-me-ln-drug-tunnel-20160421-story.html

– Synnott, Mark: 'Is This the Underground Everest?', *National Geographic*, March 2017, https://www.nationalgeographic.com/magazine/2017/03/dark-star-deepest-cave-climbing-uzbekistan/

– Thurman, Judith: 'First Impressions', *The New Yorker*, 23 June 2008, https://www.newyorker.com/magazine/2008/06/23/first-impressions

– Votintseva, Antonina et al: 'The Dark Star of Baisun-tau: A history of cave exploration in Southern Uzbekistan, 1990–2013', *Cave and Karst Science*, Transactions of the British Cave Research Association, March 2014

– Weisman, Alan: *World Without Us*, Virgin Books, London, 2008

– Wohlleben, Peter: *The Hidden Life of Trees, What They Feel, How They Communicate*, William Collins, London, 2016

↑
After nearly 2,000 years below ground, the skeletons of Herculaneum are finally being unveiled, along with the secrets that were buried with them (see page 77).

Picture Credits

Index

Page references for pictures and maps are indicated in *italics*

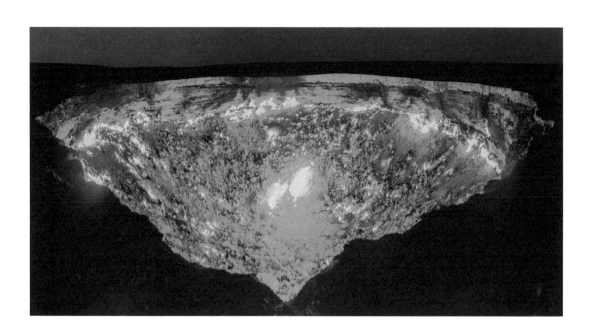

↑
The blazing inferno of Darvaza,
a sly peek into the evocative
and awe-inspiring subterranean
realm (see page 156).